THE JOY OF STATISTICS

Steve Selvin is a professor of biostatistics and epidemiology at the University of California, Berkeley. He has taught on the Berkeley campus for more than 40 years. Professor Selvin is also a member of the Johns Hopkins School of Public Health faculty and has taught in the Summer Institute of Biostatistics and Epidemiology for the last fifteen years. He lives in the Berkeley hills with two cats, one dog and a wife who is a well-known ceramic artist. He has authored or co-authored more than 250 scientific papers in the area of statistics applied to epidemiological/health issues with emphasis on birth defects and childhood cancer. In addition he has written 10 books on applied statistical methods. He has received a number of awards for teaching excellence, including the most prestigious award given by the University of California called the Berkeley Citation. His present research concerns the analysis of spatial patterns of childhood cancers in the state of California over the last decade.

T0202362

THE JOY OF STATISTICS

A Treasury of Elementary Statistical Tools
and their Applications

Steve Selvin

OXFORD
UNIVERSITY PRESS

Great Clarendon Street, Oxford, OX2 6DP,
United Kingdom

Oxford University Press is a department of the University of Oxford.
It furthers the University's objective of excellence in research, scholarship,
and education by publishing worldwide. Oxford is a registered trade mark of
Oxford University Press in the UK and in certain other countries

First published 2019
First published in paperback 2024

Published in the United States of America by Oxford University Press
198 Madison Avenue, New York, NY 10016, United States of America

British Library Cataloguing in Publication Data

Data available

Library of Congress Cataloging in Publication Data

Data available

ISBN 978–0–19–883344–4 (Hbk.)
ISBN 978–0–19–889694–4 (Pbk.)

DOI: 10.1093/oso/9780198833444.001.0001

For my grandsons Benjamin and Eli

Preface

Many introductory statistics textbooks exist for one of two purposes: as a text of statistical methods required by a variety of disciplines or courses leading to more advanced statistical methods with the goal of providing statistical tools for analysis of data. "The Joy of Statistics" is not one of these books. It is an extensive discussion of the many roles statistics plays in every day life with explanations and examples of how statistics works to explore important and sometimes unimportant questions generated from data.

A few of examples of these questions:

Who is Monty Hall?

Why is Florence Nightingale in the statistics hall of fame?

What is the relationship between a father's height, his son's height, and Sports Illustrated magazine?

How do we know the number of gray whales living in the Pacific Ocean?

How accurate are home drug testing kits?

Is 0.11 a large number?

Does a dog owner likely own a cat?

What is the difference between the law of averages and the law of large numbers?

This book is about the beauty, utility, and often simplicity of using statistics to distill messages from data. The logic and magic of statistics, without extensive technical details, is applied to answer a wide variety of questions generated from collected data. A bit of algebra, 10th grade algebra, and some elementary statistical/mathematical notation provide clear and readily accessible descriptions of a large number of ways statistics provides a path to human decision making. Included are a few classic "statistical" jokes and puzzles. Also bits of statistical history and brief biographies of important statisticians are sprinkled among various topics. The presented material is not a progression from simple to less simple to challenging techniques. The more than 40 topics present an anthology of various statistical "short stories" intended to be the first step into the world of statistical logic and methods. Perhaps the title of the text should be an "Elementary Introduction to the book Elementary Statistical Analysis."

Acknowledgments

I would especially like to thank my daughter Dr. Elizabeth Selvin, my son-in-law, David Long, and my wife of 53 years, artist Nancy Selvin for their constant support and encouragement. I owe thanks to the many colleagues and friends at the University California Berkeley and Johns Hopkins School of Public Health who have taken an enthusiastic interest in my work over the years. I also would like to acknowledge Jenny Rosen for her technical assistance and the design skills of Henna artist Robyn Jean for inspiring the cover artwork.

Contents

1

Probabilities—rules and review

> Statistics and probability, statistics and probability
> Go together like data and predictability
> This I tell you brother
> You can't have one without the other.
>
> (With apologies to FRANK SINATRA.)

Probability theory is certainly one of the most difficult areas of mathematics. However, with little effort, probabilities can be simply used to effectively summarize, explore, and analyze statistical issues generated from sampled data.

A probability is a numeric value always between zero and one used to quantify the likelihood of occurrence of a specific event or events. It measures the likelihood a specific event occurs among all possibilities. For example, roll a die and the probability the top face is a one is 1/6 or the probability it is less than 3 is 2/6 because these are two specific events generated from six equally likely possibilities. Typical notation for an event denoted A is probability $= P(A)$. In general, a probability is defined as the count of occurrences of a specific event divided by the number of all possible equally likely events. Thus, a probability becomes a statistical tool that provides a rigorous and formal assessment of the role of chance in assessing collected data.

For two events labeled A and B, a probability measures:

$P(A)$ = probability event A occurs
$P(\overline{A})$ = probability event A does not occur
$P(B)$ = probability event B occurs
$P(\overline{B})$ = probability event B does not occur.

Joint probabilities:

$P(A$ and $B)$ = probability both events A and B occur
$P(A$ and $\overline{B})$ = probability event A occurs and event B does not occur
$P(\overline{A}$ and $B)$ = probability event A does not occur and event B occurs
$P(\overline{A}$ and $\overline{B})$ = probability event A does not occur and event B does not occur

The Joy of Statistics: A Treasury of Elementary Statistical Tools and their Applications. Steve Selvin. © Steve Selvin 2019. Published in 2019 by Oxford University Press. DOI: 10.1093/oso/9780198833444.001.0001

Conditional probabilities:

$P(A \mid B)$ = probability event A occurs when event B has occurred or
$P(B \mid A)$ = probability event B occurs when event A has occurred.

Note: the probability $P(A) = 0$ means event A cannot occur ("impossible event"). For example, rolling a value more than six with a single die or being kidnaped by Martians. Also note: the probability $P(A) = 1$ means event A always occurs ("sure event"). For example, rolling a value less than 7 with a single die or not being kidnaped by Martians.

A repeat of joint probabilities for two events A and B usefully displayed in a 2×2 table:

Table 1.1 Joint distribution of events A and B

events	B	$\overline{B}$	sum
A	$P(A$ and $B)$	$P(A$ and $\overline{B})$	$P(A)$
$\overline{A}$	$P(\overline{A}$ and $B)$	$P(\overline{A}$ and $\overline{B})$	$P(\overline{A})$
sum	$P(B)$	$P(\overline{B})$	1.0

A subset of data ($n = 100$ observations) from a national survey of pet ownership conducted by the US Bureau of Labor Statistics containing counts of the number of people surveyed who own at least one dog (event denoted D) or own at least one cat (event denoted C) or own neither a dog or a cat (event denoted $\overline{D}$ and $\overline{C}$) or own both a cat and a dog (event denoted D and C).

Table 1.2 Joint distribution of 100 cat and dog owners

	D	$\overline{D}$	sum
C	D and $C = 15$	$\overline{D}$ and $C = 45$	$C = 60$
$\overline{C}$	D and $\overline{C} = 20$	$\overline{D}$ and $\overline{C} = 20$	$\overline{C} = 40$
sum	$D = 35$	$\overline{D} = 65$	$n = 100$

Some specific cat and dog probabilities

$P(D) = 35/100 = 0.35$ and $P(C) = 60/100 = 0.60$
$P(D$ and $C) = 15/100 = 0.15$, $P(\overline{D}$ and $C) = 45/100 = 0.45$
$P(D$ and $\overline{C}) = 20/100 = 0.20$, $P(\overline{D}$ and $\overline{C}) = 20/100 = 0.20$
$P(D$ or $C) = P(D$ and $C) + P(\overline{D}$ and $C) + P(D$ and $\overline{C}) = 0.15 + 0.45$
$\quad\quad + 0.20 = 0.80$

also, $P(D \text{ or } C) = 1 - P(\overline{D} \text{ and } \overline{C}) = 1 - 0.20 = 0.80$ because
$P(D \text{ or } C) + P(\overline{D} \text{ and } \overline{C}) = 1.0$.

Conditional probabilities

Probability a cat owner owns a dog $= P(D \,|\, C) = 15/60 = 0.25$ (row: condition $=$ cat $= C$);
probability a dog owner owns a cat $= P(C \,|\, D) = 15/35 = 0.43$ (column: condition $=$ dog $= D$).

In general, for events A and B, then

$$P(A\,|\,B) = \frac{P(A \text{ and } B)}{P(B)} - - \text{condition} = B$$

and

$$P(B\,|\,A) = \frac{P(A \text{ and } B)}{P(A)} - - \text{condition} = A.$$

For example, again dogs and cats:

$$P(D\,|\,C) = \frac{P(D \text{ and } C)}{P(C)} = \frac{15/100}{60/100} = 0.25 - - \text{condition} = \text{cat}$$

and

$$P(C\,|\,D) = \frac{P(D \text{ and } C)}{P(D)} = \frac{15/100}{35/100} = 0.43 - - \text{condition} = \text{dog}.$$

Independence

Two events A and B are independent when occurrence of event A is unrelated to occurrence of event B. In other words, occurrence of event A is not influenced by occurrence of event B and vice versa.

Examples of independent events:

toss coin: first toss of a coin does not influence the second toss,
birth: boy infant born first does not influence the sex of a second child,
lottery: this week's failure to win the lottery does not influence winning next week,
politics: being left-handed does not influence political party affiliation, and
genetics: being a male does not influence blood type.

Notation indicating independence of events A and B is

$$P(A|B) = P(A) \quad \text{or} \quad P(B|A) = P(B).$$

Thus, from the pet survey data, the conditional probability $P(D \mid C) = 0.25$ is not equal to $P(D) = 0.35$, indicating, as might be suspected, that owning a dog and a cat are not independent events. Similarly, $P(C \mid D) = 0.43$ is not equal to $P(C) = 0.60$, necessarily indicating the same dog/cat association.

An important relationship:

$$P(A|B) = \frac{P(A \text{ and } B)}{P(B)}.$$

Furthermore, $P(A \mid B) \times P(B) = P(A) \times P(B) = P(A \text{ and } B)$ when events A and B are independent because then $P(A \mid B) = P(A)$. Incidentally, expression $P(A \mid B) \times P(B) = P(A \text{ and } B)$ is called the *multiplication rule*. In addition, when events $\{A, B, C, D, \ldots\}$ are independent, joint occurrence:

$$P(A \text{ and } B \text{ and } C \text{ and } D \text{ and } \cdots) = P(A) \times P(B) \times P(C) \times P(D) \times \cdots.$$

Statistics pays off

Capitalizing on lack of independence in a gambling game called black-jack made Professor Edward Thorp famous, at least in Las Vegas.

First a short and not very complete description of the rules of this popular casino card game that requires only a standard deck of 52 playing cards.

The object is to beat the casino dealer by:

1. a count higher than the dealer without exceeding a score of 21 or
2. the dealer drawing cards creating a total count that exceeds 21 or
3. a player's first two cards are ace and a ten count card.

The card counts are face values 2 through 10, jack, queen, and king count 10, and ace counts one or eleven.

At the casino gaming table, two cards are dealt to each player and two cards to the dealer. Both players and dealer then have options of receiving additional cards. The sum determines the winner. The dealer starts a new game by dealing a second set of cards using the remaining cards in the deck. Thus inducing a dependency (lack of independence) between cards already played and cards remaining in the deck. Professor Thorp (1962) realized the house advantage could be overcome by simply counting the number of 10-count cards played in the previous game. Thus, when the remaining deck contains a large number of 10-count cards it gives an advantage to the player and when the cards remaining in the deck lack 10-count cards the advantage goes to the dealer. Professor Thorp simply bet large amounts of money when the deck was in his favor (lots of 10-count cards) and small amounts when the deck was not in his favor (few 10-count cards), yielding a small but profitable advantage. That is, cards left for the second game depend on the cards played in the first game, producing non-independent events yielding a detectable pattern.

Blackjack rules were immediately changed (using eight decks of cards, not one, for example) making card counting useless. Thus, like casino games roulette, slots machines, keno, and wheel of fortune, blackjack also does not produce a detectable pattern. That is, each card dealt is essentially independent of previous cards dealt.

The playing cards could be shuffled after each game producing independence, but this would be time consuming, causing a monetary loss to the casino. Also of note, Professor Thorp wrote a book about his entire experience entitled "Beat the Dealer."

Picture of Probabilities—Events *A* and *B* (circles)

$$P(A) = \frac{9}{20} = 0.45 \quad P(B) = \frac{7}{20} = 0.35$$

$$P(A \text{ and } B) = \frac{3}{20} = 0.15 \quad P(A \text{ or } B) = \frac{12}{20} = 0.6 \quad P(\overline{A} \text{ and } \overline{B}) = \frac{7}{20} = 0.35$$

$$P(A \mid B) = \frac{3}{7} = 0.43 \quad P(B \mid A) = \frac{3}{9} = 0.33$$

Table of the same data—an association

	B	not B	total
A	3	6	9
not A	4	7	11
total	7	13	20

Picture of Probabilities—Independent Events *A* and *B* (circles)

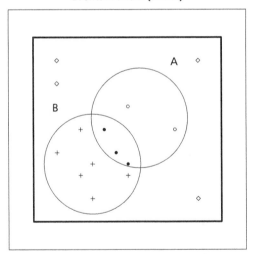

$$P(A) = \frac{5}{15} = 0.33 \quad P(B) = \frac{9}{15} = 0.60$$

$$P(A \text{ and } B) = \frac{3}{15} = 0.20 \quad P(A \text{ or } B) = \frac{11}{15} = 0.73 \quad P(\overline{A} \text{ and } \overline{B}) = \frac{4}{15} = 0.27$$

Table of the same data—no association

	B	not B	total
A	3	2	5
not A	6	4	10
total	9	6	15

Probability of event A restricted to occurrence of event B—again denoted $P(A \mid B)$ is:

$$P(A \mid B) = \frac{3}{9} = 0.33 - \text{—column B.}$$

The probability of event A is not influenced by event B, both $P(A)$ and $P(A \mid B)$ equal 0.33. In symbols, $P(A \mid B) = P(A)$. Technically, the two events are said to be independent. More technically, they are said to be stochastically independent. Also, necessarily $P(B \mid A) = P(B) = 0.60$. An important consequence of independence of two events, as noted, is $P(A$ and $B) = P(A) \times P(B)$. Specifically, from the example, $P(A$ and $B) = 0.20$ and, therefore, $P(A) \times P(B) = (0.33)(0.60) = 0.20$.

These two-circle representations of joint probabilities are called Venn diagrams (created by John Venn, 1880).

Roulette

A sometimes suggested strategy for winning at roulette:

Wait until three red numbers occur then bet on black.

Red (R) and black (B) numbers appear with equal probabilities or $P(B) = P(R)$. Let R_1, R_2, and R_3 represent occurrence of three independent red outcomes and B represents occurrence of an additional independent black outcome. Then, consecutive occurrences of three red outcomes followed by a single black outcome dictates that:

$$P(B \mid R_1 R_2 R_3) = \frac{P(B \text{ and } R_1 R_2 R_3)}{P(R_1 R_2 R_3)}$$

$$= \frac{P(B)P(R_1)P(R_2)P(R_3)}{P(R_1)(R_2)(R_3)} = P(B).$$

No change in the probability of the occurrence of black! The red outcomes must be unpredictable (independent) or, for example, everyone would bet on black after red occurred, or vice versa, making the game rather boring. If a successful strategy existed, the casino game of roulette would not.

One last note:

The game of roulette was originated by the famous 17th century mathematician/physicist Blaise Pascal and the word roulette means little wheel in French.

Summation notation (Σ)

Data:

$$\{x_1 = 1, x_2 = 2, x_3 = 3, x_4 = 4, x_5 = 5, x_6 = 6, x_7 = 7\}$$

and number of observations $= n = 7$.

Sum (denoted S):

$$S = x_1 + x_2 + x_3 + x_4 + x_5 + x_6 + x_7 = 1 + 2 + 3 + 4 + 5 + 6 + 7 = 28$$
$$\text{denoted} \quad \Sigma x_i = 28.$$

Mean value (denoted $\overline{x}$), then:

$$\overline{x} = \frac{S}{n} = \frac{x_1 + x_2 + x_3 + x_4 + x_5 + x_6 + x_7}{n} = \frac{1 + 2 + 3 + 4 + 5 + 6 + 7}{7}$$
$$= \frac{28}{7} = \frac{\Sigma x_i}{7} = 4.$$

Sum of squared values $(x_i - \overline{x})^2$:

$$\Sigma(x_i - \overline{x})^2 = (x_1 - \overline{x})^2 + (x_2 - \overline{x})^2 + (x_3 - \overline{x})^2 + (x_4 - \overline{x})^2 + (x_5 - \overline{x})^2$$
$$+ (x_6 - \overline{x})^2 + (x_7 - \overline{x})^2$$
$$= (1 - 4)^2 + (2 - 4)^2 + (3 - 4)^2 + (4 - 4)^2 + (5 - 4)^2$$
$$+ (6 - 4)^2 + (7 - 4)^2 = 28$$

and more details

$$\Sigma(x_i - \overline{x})^2 = (-3)^2 + (-2)^2 + (-1)^2 + (0)^2 + (1)^2 + (2)^2 + (3)^2$$
$$= 9 + 4 + 1 + 0 + 1 + 4 + 9 = 28.$$

If

$$\{x_1 = 3, x_2 = 2, x_3 = 1\} \text{ and } \{y_1 = 1, y_2 = 2, y_3 = 3\}$$

then the mean values are $\bar{x} = \bar{y} = 2$ and

$$\sum x_i y_i = 3(1) + 2(2) + 1(3) = 10.$$

Application:

$$correlation\,coefficient = r = \frac{\sum(x_i - \bar{x})(y_i - \bar{y})}{\sqrt{\left(\sum(x_i - \bar{x})^2 \times \sum(y_i - \bar{y})^2\right)}}$$

$$r = \frac{(3-2)(1-2) + (2-2)(2-2) + (1-2)(3-2)}{\sqrt{\left(\left[(3-2)^2 + (2-2)^2 + (1-2)^2\right] \times \left[(1-2)^2 + (2-2)^2 + (3-2)^2\right]\right)}}$$

$$r = \frac{(1)(-1) + (0)(0) + (-1)(1)}{\sqrt{\left([1^2 + 0 + 1^2] \times [1^2 + 0 + 1^2]\right)}} = \frac{-2}{\sqrt{2 \times 2}} = \frac{-2}{2} = -1.0$$

Types of variables

Qualitative:

types	examples		
nominal	socioeconomic status	ethnicity	occupation
ordinal	educational levels	military ranks	egg sizes
discrete	counts	reported ages	cigarettes smoked
binary	yes/no	exposed/unexposed	case/control

Quantitative:

types	examples		
continuous	weight	distance	time
ratio	speed	rate	odds

Reference: *Beat the Dealer*, by Edward O. Thorp, Vintage Books, 1966

2

Distributions of data—four plots

Two quotes from the book entitled Gadsby by Ernest Vincent Wright:

First and last paragraphs:

If youth, throughout all history, had a champion to stand up for it; to show a doubting world that a child can think; and, possibly, do it practically; you would not constantly run across folks today who claim 'a child do not know anything.' A child's brain starts functioning at birth; and has amongst its many infants convolutions, thousands of dormant atoms, into which God has put a mystic possibility for noticing an adult acts, and figuring out it purport.

A glorious full moon sails across a sky without a cloud. A crisp night air has folks turning up coats collars and kids hopping up and down for warmth. And that giant star, Sirius, winking slily, knows that soon, that light up in his honors room window will go out. Fttt! It is out! So, as Sirius and Luna hold an all night vigil, I will say soft 'Good-night' to all our happy bunch, and to John Gadsby, youth's champion.

Question: Notice anything strange?

A table and plot of letter frequencies from the first and last paragraphs from the Gadsby book clearly show the absence of the letter "e." A table or a plot makes the absence of letter "e" obvious. In fact, the entire book of close to 50,000 words does not contain letter "e." Ironically, the author's name contains the letter "e" three times. The plotted distribution of indeed an extreme example illustrates the often effective use of a table or simple plot to identify properties of collected data.

Table 2.1 Distribution of letters (counts)

a	b	c	d	e	f	g	h	i	j	k	l	m
68	7	22	30	0	15	22	39	61	1	12	37	1

n	o	p	q	r	s	t	u	v	w	x	y	z
10	60	74	16	28	55	67	32	2	15	18	18	1

The Joy of Statistics: A Treasury of Elementary Statistical Tools and their Applications. Steve Selvin. © Steve Selvin 2019. Published in 2019 by Oxford University Press. DOI: 10.1093/oso/9780198833444.001.0001

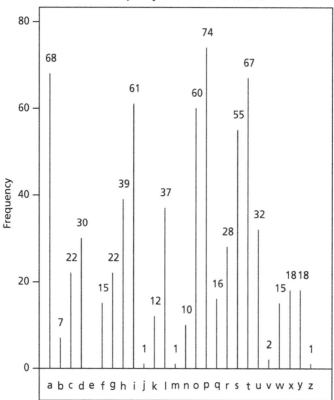

Frequency distribution of letters

Frequency plots are basic to describing many kinds of data. A large number of choices exist among graphical representations. Four popular choices are: a barplot, a histogram, a stem-leaf plot, and a frequency polygon.

Barplot

The height of each bar indicates the frequency of the values of the variable displayed. Order of the bars is not a statistical issue. The example is a comparison of the number of world champion chess players from

Chess champions

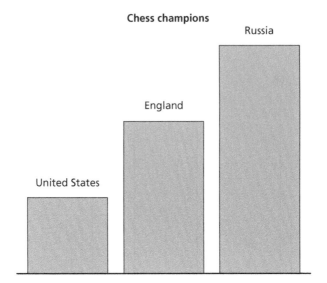

each country (height = frequency). For example, visually, Russia has produced three times more world champions than the US.

Histogram

A histogram is distinctly different. Sample data ($n = 19$), ordered for convenience:

$$X = \{21, 24, 27, 29, 42, 44, 48, 67, 68, 71, 73, 78, 82, 84, 86, 91, 95, 96, 99\}.$$

A histogram starts with an ordered sequence of numerical intervals. In addition, a series of rectangles are constructed to represent each of the frequencies of the sampled values within each of the sequence of these intervals.

The example displays frequencies of observed values labeled X classified into four intervals of size 20. That is, four intervals each containing 4, 3, 5, and 7 values. The areas of the rectangles again provide a direct visual comparison of the data frequencies. The resulting plot, like the previous plots, is a visual description of the distribution of collected numeric values.

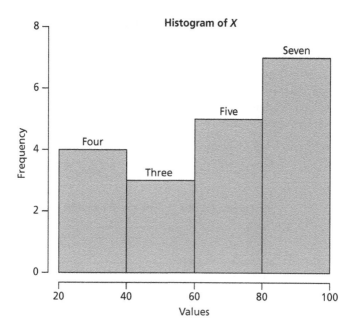

Stem-leaf plot

The stem-leaf plot is another approach to visually displaying data but often differs little from a histogram. The "stem" is the leftmost column of the plot usually creating single digit categories ordered smallest to largest. Multiple digit categories can be also used to create the stem. This "stem" is separated from the "leaves" by a vertical bar {" | "}. The "leaves" are an ordered list creating rows made up of the remaining digits from each category. That is, the stem consists of an ordered series of initial digits from observed values and leaves consist of remaining digits listed to the right of their respective stem value. For example, the stem-leaf plot of the previous data labeled X for intervals {20, 40, 60, 80, 100} is:

stem	leaves
2	1479
4	248
6	78138
8	2461569

The previous histogram and this stem-leaf plot hardly differ in principle. If the intervals selected to construct a histogram correspond to those of the stem, the stem-leaf plot is essentially a histogram rotated ninety degrees, illustrated by the previous example data (X). Unlike a histogram, the original data values are not lost. Gertrude Stein famously said, "A difference to be a difference must make a difference."

Another example, ordered sample data ($n = 20$):

$$X = \{0, 1, 2, 3, 4, 5, 11, 12, 13, 14, 15, 21, 22, 23, 24, 31, 32, 33, 41, 42\}$$

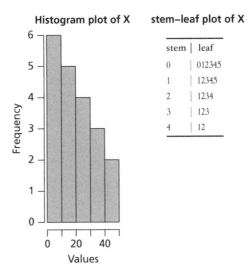

Frequency polygon

The frequency polygon is a close cousin of the histogram. It is no more than a series of straight lines that result from joining the midpoints of the tops of the histogram rectangles. Like a histogram, it displays a comparison of data frequencies as well as the general shape of the distribution of collected values. A sometimes useful difference between a histogram and a frequency polygon is several frequency polygons can be displayed on the same set of axes.

Data ($n = 30$):

$$X = \{-2.03, -1.88, -1.74, -1.49, -0.78, -0.54, -0.54, -0.40, -0.40,$$
$$-0.38, -0.30, -0.21, -0.03, -0.03, 0.05, 0.20, 0.23, 0.43, 0.49, 0.50,$$
$$0.51, 0.62, 0.67, 0.96, 0.97, 1.11, 1.27, 1.44, 1.64, 2.31\}$$

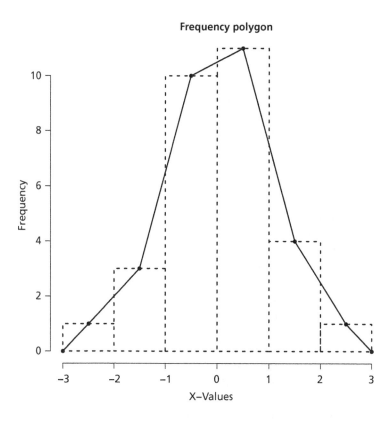

Frequency polygon

Reference: *Gadsby*, by Ernest Vincent Wright, Wetzel Publishing Co, 1919
Note: the full text is available on the Internet—Internet archive.

3

Mean value—estimation and a few properties

The sum of a series of sampled values (each value denoted x_i) divided by the number of values sampled (denoted n) is almost always denoted $\bar{x}$. In symbols, for data values represented by $\{x_1, x_2, x_3, \cdots, x_n\}$, then:

$$estimated\ mean\ value = \bar{x} = \frac{x_1 + x_2 + x_3 + \cdots + x_n}{n} = \frac{1}{n}\sum x_i$$

is a summary value formally called the arithmetic mean. Sometimes carelessly called the average. It is a specific kind of average. A number of statistical measures are also special kinds of averages. A few are: median, mode, range, harmonic mean, geometric mean, and midrange.

An estimated mean value is most meaningful and easily interpreted when collected data are sampled from a symmetric distribution. It is then that the estimated mean value locates the middle of the sampled values, sometimes called the most "typical" value and usually considered the "best" single value to characterize an entire set of observations. Physicists would call the mean value the "center of gravity." Others call it a measure of central tendency or a measure of location. When the estimated mean value is subtracted from each of n observations, the sum of these differences is zero (in symbols, $\sum(x_i - \bar{x}) = 0$), as would be expected of an estimate at the center of a sample of observations.

Also important, when a value is again subtracted from each of the observed values x_i, the sum of these differences squared has the smallest possible value if the value subtracted is the estimated mean value $\bar{x}$. Thus, the sum of squared deviations from the mean value (in symbols, $\sum(x_i - \bar{x})^2$) is minimized by choice of $\bar{x}$. In this sense, estimate $\bar{x}$ is the single value "closest" to the values sampled. Any value other than the mean value produces a larger sum of squared deviations.

When sampled values are not symmetrically distributed, the mean value is usually not a useful single value characterization of collected data. The median value is often suggested as a substitute.

The Joy of Statistics: A Treasury of Elementary Statistical Tools and their Applications. Steve Selvin. © Steve Selvin 2019. Published in 2019 by Oxford University Press. DOI: 10.1093/oso/9780198833444.001.0001

Formal dictionary definition of a median value:

> the value in an ordered set of data such that the number of observations below and above are equal in number or the value denoted m such that there is an equal probability of an observation falling below or above the value m.

Thus, the median value also describes location or central tendency and, in addition, it is not disproportionally dominated by large or small extreme values.

Consider artificial data labeled X and its distribution.

The data: $X = \{2, 5, 9, 20, 35, 60, 95\}$ yield the not very representative mean value $\bar{x} = 32.3$.

additive scale = as measured

The logarithms of these seven values have mean value $\overline{log(x)} = 2.83$, which is a better single characterization of the data X.

logarithim scale = log–values

A comparison of the distribution of example values (X) to logarithms of the same values ($log[X]$) begins to indicate why mean values and other summary statistics are sometimes estimated from logarithms of observed values.

Logarithms produce small reductions in small values relative to considerably larger reductions in large values. From the example, $log(2) \rightarrow$ 0.7 and $log(95) \rightarrow 4.6$. Thus, log-values tend to create more symmetric distributions when the data themselves have an asymmetric distribution. The example data X illustrate this key property. Using a log-value distribution to estimate the mean value can yield a more representative statistical summary because logarithms of observed values mute influences of extreme large and small asymmetric values and, as illustrated, tend to produce a more symmetric distribution.

A variety of advantages exist from employing a logarithmic scale. A number of statistical techniques require data to be sampled from at least approximately symmetric distributions, which can often be generated from a logarithmic transformation. A log-transformation can reveal relationships not obvious when the data have an asymmetric distribution. For example, two parallel lines on a logarithmic scale indicate a constant difference on an additive scale (examples follow). Occasionally, logarithms of observed values reveal useful summaries such as linear or additive relationships. In addition, details sometimes lost, such as in a graphical display containing extremely small values, become apparent when displayed on a log-scale.

An important application of an estimated mean value arises in measuring variation among sampled observations. The essence of statistical analysis is often the description of collected data accounting for influences that cause distracting variation.

For a sample of n values, as before, represented as $\{x_1, x_2, x_3, \cdots, x_n\}$, the mean of the observed squared deviations from the mean value $\overline{x}$ measures variation. The deviations again are $x_i - \overline{x}$ and the mean of the squared values, estimated in the usual way (the sum divided by n), becomes:

$$mean\,squared\,deviation = \frac{1}{n}\left[(x_1 - \overline{x})^2 + (x_2 - \overline{x})^2 + (x_3 - \overline{x})^2 + \cdots + (x_n - \overline{x})^2\right]$$

$$= \frac{1}{n}\sum (x_i - \overline{x})^2.$$

That is, the mean squared deviation is exactly what the name suggests. It is simply the mean of the squared deviations directly measuring the extent of variation among sampled values.

Another popular statistic used to describe variation among sampled values is the same sum of squared deviations but divided by $n-1$ and not n where:

$$sample\,variance = S_X^2 = \frac{1}{n-1}\left[(x_1 - \overline{x})^2 + (x_2 - \overline{x})^2 + (x_3 - \overline{x})^2 + \cdots + (x_n - \overline{x})^2\right]$$

$$= \frac{1}{n-1}\sum (x_i - \overline{x})^2.$$

The value produced is a slightly more accurate estimate of the variability of the distribution sampled and is called the *sample variance* (symbol S_X^2). The square root is called the *standard deviation* (symbol S_X).

Examples of sample variability

Three sets of artificial data illustrate three summary statistics, particularly variation measured by sample variance and mean squared deviation:

$X = \{1, 2, 3, 4, 5, 6, 7, 8, 9, 10, 11, 12, 13\}$
$n = 13$, mean value $= 7$, sample variance$(X) = 15.17$, mean squared deviation $= 14.0$
$Y = \{1, 3, 5, 7, 9, 11, 13\}$
$n = 7$, mean value $= 7$, sample variance$(Y) = 18.67$, mean squared deviation $= 16.0$
$Z = \{1, 4, 7, 10, 13\}$
$n = 5$, mean value $= 7$, sample variance$(Z) = 22.50$, mean squared deviation $= 18.0$.

Increasing sample variances and mean squared deviations reflect increasing variability within each of three data sets. Clearly, the difference between these two measures of variability is small for most sample sizes $(1/(n-1) \approx 1/n)$. For $n = 10$, the difference is $0.11-0.10 = 0.01$.

A random sample from a distribution with mean of 7 and variance of 14:

data $= D = \{5.24, 2.73, 8.06, 8.57, 12.30, 6.06, 13.25, 5.49, 9.13, 6.66, 13.04, 1.30, 3.15\}$ $n = 13$, mean value $= 7.3$, sample variance$(D) = 15.1$, mean squared deviation $= 14.0$.

Law of averages

The law of averages, sometimes confused with the law of large numbers, is not a law but a belief. It is the belief that random events, like tossing a coin, tend to even out over even a short run. That is, if a coin is tossed ten times and heads appears on all ten tosses, the law of averages states the likelihood of tails occurring on the eleventh toss has increased. If events, like tossing a coin, are independent, then past events do not influence current or future events. Thus, the law of averages has earned the name "gambler's fallacy" or "gambler's fantasy." Nevertheless, the law of averages continues to attract a large number of faithful believers.

A casino in Monte Carlo, France established a record of sorts. On a roulette wheel, a common bet is on the occurrence of black or red numbers that are equally likely (probability, red $=$ black). On a particular evening in 1903, red occurred consecutively 39 times. If one dollar was bet on the first red and cashed-in exactly on the 39th occurrence of red, the winnings would have accumulated to more than 10 billion dollars. Consistent with the law of averages, a frenzy of betting on black occurred with higher and higher stakes as more and more consecutive red numbers occurred. It is said at least a million dollars was lost (worth today 27.8 million). Incidentally, the American roulette record of 26 consecutive black outcomes occurred in an Atlantic City casino in 1943.

When a coin is tossed and 10 consecutive heads occurs, a mathematician says "Wow, a rare event" and a statistician says "Let me see the coin."

Law of large numbers

The law of large numbers is definitely a law and important not only in the context of statistics but critical in a huge variety of situations. The law of large numbers states that as the number of observations increases, a value estimated from these observations converges to an essentially stable value. For example, it is highly unlikely an individual knows or even can accurately guess the likelihood of surviving beyond age 65. However, large quantities of data gathered on length of life produce an extremely precise estimate of this probability.

The price of life insurance can not be determined without the law of large numbers. Transit systems, with great precision, depend on knowing how many travelers will arrive at 8:00 Monday mornings. Budgets are created from precise estimates based on expected tax returns. If the law of large numbers was repealed, book publishers would not know how many books to print, social planning of all sorts would be chaotic and car manufacturers would not know how many pick-up trucks to assemble. Theater, restaurant, and airline attendance would fluctuate unpredictably from none to over capacity. Precise predictability from data is guaranteed by the law of large numbers, a necessary component of modern society.

The mean value simply illustrates the law of large numbers. As the sample size (n) increases, precision of an estimated value increases. In other words, the variation of an estimated value decreases, producing a stable and ultimately an extremely precise and known value. In the case of an estimated mean value, variability is directly expressed as:

$$variance(\overline{x}) = \frac{variance\, of\, x}{n}.$$

Formal expression of the variance of the estimated mean value clearly shows the larger the sample, the more precise the estimated value. Thus, for large samples of data, the estimated mean value becomes essentially the value estimated because its variability becomes essentially zero.

If μ represents the unknown mean value, then as n increases, the estimated mean value becomes increasingly indistinguishable from μ because its variance becomes essentially zero and thus, in symbols, $\overline{x} \rightarrow \mu$. That is, the law of large numbers guarantees a precise, stable, and predictable estimated value. Like the mean value, most estimates increase in precision with increasing sample size.

One last note

A recent survey indicated that 65% of Americans believe they have above average intelligence. That is, the average American thinks they are smarter than the average American.

4

Boxplots—construction and interpretation

Sometimes a cliche tells the whole story—"A picture is worth a thousand words."

A boxplot is a graphical display of several specific properties of a distribution of observed values. The box part of a boxplot is constructed so the bottom of the box is the location of the lower quartile (first quartile = 25%) and the top of the box is the location of the upper quartile (third quartile = 75%) of collected data. Thus, the size of the box represents the interquartile range. That is, the height of the box part of the boxplot represents the location of 50% of the sample data reflecting variability/range. The width of the box is usually made proportional to sample size, which is only an issue when boxplots are compared. The box is divided by a line at the median value giving a sense of data location. The distance from the median line to the top of the box relative to the distance to the bottom reflects symmetry of the sampled distribution. A line exactly at the middle identifies an exactly symmetric distribution. Otherwise, comparison of the distances above and below the median line visually indicates extent and direction of asymmetry. Lines extended from top and bottom of the box, each of length 1.5 times the interquartile range, are called "whiskers." Observations beyond these limits are plotted individually (circles). It is frequently important to identify extreme values likely to produce strong influences on summary statistics or occasionally locate observations not belonging in the collected data. An "outlier value" is sometimes defined as an observation that lies beyond whisker length.

Another feature of a boxplot is that side-by-side display of two or more boxplots creates a visual comparison of several different sampled distributions. For example, median values display differences in locations of each distribution and sizes of the compared boxes potentially identify differences in sample size, symmetry, and variability.

The Joy of Statistics: A Treasury of Elementary Statistical Tools and their Applications. Steve Selvin. © Steve Selvin 2019. Published in 2019 by Oxford University Press. DOI: 10.1093/oso/9780198833444.001.0001

The following are applications of boxplot representations.

The properties of a typical box plot

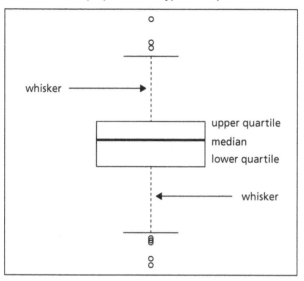

Histogram of data

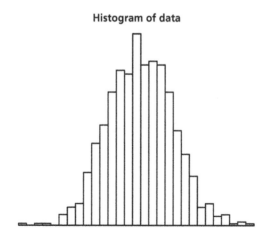

Boxplot of same data

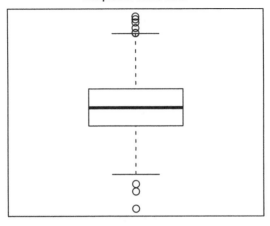

Histogram of data

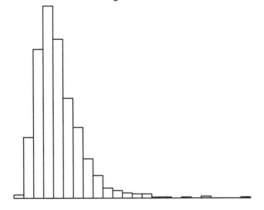

Boxplot of same data

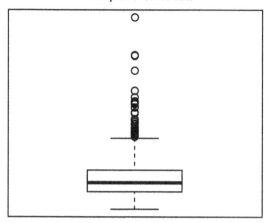

Boxplots of birth weight by parity

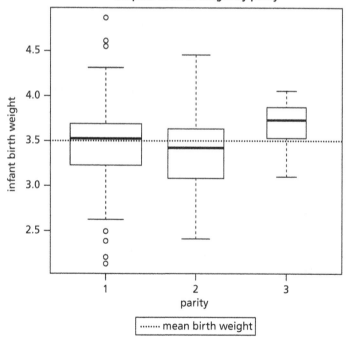

Decreasing widths of each box indicate a corresponding decrease in sample size.

5

The lady who tasted tea—a bit of statistical history

In the beginning, on a summer afternoon in Cambridge, England (circa 1920), an extraordinary meeting took place between two totally different people: a British lady who bravely claimed she could tell whether tea was prepared by adding milk to tea or tea to milk and Ronald Aylmer Fisher who was destined to become one of the world's most renowned statisticians.

The lady's claim was met with skepticism. Clearly tasting one cup of tea would not provide satisfactory evidence that the "tea-tasting lady" had an unusual ability. In the complete absence of an ability to determine two different methods of preparation her answer would be correct half the time. Fisher proposed an experiment.

History does not record the details but it was decided eight cups of tea would be prepared, four with tea added to milk and four with milk added to tea, producing perhaps the most famous 2 × 2 table in the history of statistics:

Table 5.1 Tea-tasting lady

	"tea to milk"	"milk to tea"	total
tea added to milk	4	0	4
milk added to tea	0	4	4

The results from Fisher's experiment produced substantial but intuitive evidence that at least one person could tell the difference between these two tea preparations. This Sunday afternoon experiment remarkably crystallized two concepts fundamental to modern statistics.

First, the probability of correctly identifying all eight cups of tea by chance alone (guessing) was recognized to be a useful number. It is:

$$P(all\,eight\,correct \mid guessing) = \frac{4}{8} \times \frac{3}{7} \times \frac{2}{6} \times \frac{1}{5} = \frac{24}{1680} = 0.014.$$

The Joy of Statistics: A Treasury of Elementary Statistical Tools and their Applications. Steve Selvin. © Steve Selvin 2019. Published in 2019 by Oxford University Press. DOI: 10.1093/oso/9780198833444.001.0001

That is, the obviously small probability 0.014 indicates guessing four cups correctly is unlikely, very unlikely. Therefore, the alternative of not guessing becomes likely. Only a small doubt remains that the four correct determinations were other than a result of exceptional sense of taste.

More generally, a small probability that an observed outcome occurred by chance alone provides evidence of a systematic (nonrandom) outcome. This probability is frequently called the *level of significance* (nicknamed, *p*-value). A level of significance is a quantitative measure of circumstantial evidence. A small level of significance, such as 0.014, leads to the inference that observed results are not likely due to chance. In other words, the smaller the significance probability, the less chance is a plausible explanation.

The second concept fundamental to modern statistics that arose that afternoon was designing an experiment to answer specific questions based on a statistical analysis of resulting data. R. A. Fisher wrote an important and famous book on the topic entitled *Design of Experiments* (1935). His book contains a complete description of the tea-tasting experiment and analysis. Of more importance, this book introduced statistical methods that opened the door to creating, collecting, and analyzing experiment-generated data. Fisher provided many detailed descriptions of experimental designs that produced summary values created to be evaluated with statistical analyses. These statistical tools swept through science, medicine, agriculture, and numerous other fields for the next fifty or more years. It has been said R. A. Fisher created statistical analysis and following statisticians added details.

A bit more about R. A. Fisher

R. A. Fisher graduated from Cambridge University. He accepted a position (1919) at Rothamstead Agriculture Experiment Station (a British government research institution). He began his academic career at University College, London where he was appointed Galton Professor of Eugenics. Twenty years later he accepted a position as Professor of Genetics at Cambridge University. His numerous contributions both to theory and practice of statistics and his equally important contributions to extending the understanding of genetics have been recognized by numerous awards and honorary appointments. He was made a Knight Bachelor by Queen Elizabeth II in 1952.

Fisher's scientific papers, presentations, and books contain endless insightful suggestions and remarks. A few are:

"To call in the statistician after the experiment is done may be no more than asking him to perform a post-mortem examination: he may be able to say what the experiment died of."

"The million, million, million to one chance happens one in a million, million, million times no matter how surprised we may be at the results."

"The more highly adapted an organism becomes, the less adaptable it is to new change."

An aside: Fisher, who did not always get along with his colleagues, may have been referring to these individuals.

Maybe his most important quote:

"Natural selection is a mechanism for generating an exceedingly high degree of improbability."

A bit about R. A. Fisher's colleague, statistician Karl Pearson

Carl Pearson was born in 1857 and died in 1936. As an adult, he changed his name from Carl to Karl to honor Karl Marx. Pearson was home schooled, then attended University College, London. In 1875 he won a scholarship to King's College, London. He graduated from Cambridge University in 1879. His interests ranged over a huge variety of topics; ethics, Christianity, elasticity, applied mathematics, heredity, and evolution, to name a few. He even managed a degree in law and was admitted to the bar.

One the greatest of all statisticians, he did not begin the study of statistics until he was 33 years old (1890). Like R. A. Fisher, many of his contributions are the foundation of much of modern statistics.

A few notable contributions:

Properties of the correlation coefficient.
Method of moments estimation.
Recognition of the importance of probability distributions.
Perhaps, originator of the histogram.
Pearson's goodness-of-fit statistic was rated by *Science* magazine in the top ten scientific achievements of the twentieth century (ranked number seven in importance).

R. A. Fisher and Karl Pearson were colleagues but not friends. Pearson was co-founder and editor of the journal *Biometrika* (1901) which exists today. Fisher submitted a number of papers that were published. After much discussion and suggestions back and forth, Pearson did not publish a specific paper submitted by Fisher. Finally, Pearson relented and published Fisher's work but as a footnote. From that time on, Fisher sent his papers to less pre-eminent journals. However, years later (1945) R. A. Fisher was asked to contribute an article on Karl Pearson to the *The Dictionary of National Biography*. Fisher replied, "I will set to work and do what I can. I did not always get on very well with the subject but certainly I have read with more attention than most have been able to give him." The editor replied, somewhat mysteriously, "I am sure you will roar like any sucking dove"! (L. G. Wickham Legg).

Reference: *The Lady Tasting Tea, How Statistics Revolutionized Science in the Twentieth Century* by David Salsburg, Owl Books, 2002.

6

Outlier/extreme values
—a difficult decision

A dictionary definition of an outlier observation from an introductory statistics textbook:

> An observation that is well outside of the expected range of values in a study or experiment, and which is often discarded from the data set.

A problem with this definition from a statistical point of view is the ambiguity of the phrase "well outside." In addition, an "expected range" is typically not possible to unequivocally establish. Furthermore, a proven outlier value should always be discarded.

Two different kinds of extreme values are encountered. One is where substantial evidence arises that the observation does not belong to the collected data. For example, a maternal weight was recorded as 312 pounds. Reviewing the original data, it was discovered the correct value was in fact 132 pounds. The first maternal weight is a true out-and-out outlier. Usually, it is not that simple.

The other kind of "outlier" value occurs when data contain unusually large values that may have occurred by chance alone. The decision to include or exclude this type of outlier/extreme observation from an analysis is rarely obvious. Because the value in question is necessarily extreme, the decision can be critical.

Three, not always effective, strategies are sometimes suggested. In the presence of a possible outlier, the kind of statistical summary chosen can lessen its influence. The median value is such a summary. The median of values of $\{1, 2, 3, 4, 5\}$ and $\{1, 2, 3, 4, 50\}$ are the same. Statistical methods exist to reduce influences of outlier/extreme values. Rank procedures and other methods called nonparametric are also such techniques. These methods typically rely on replacing observed values with substitute values, such as numeric ranks or plus/minus signs, minimizing the sometimes overwhelming influence of one or

The Joy of Statistics: A Treasury of Elementary Statistical Tools and their Applications. Steve Selvin. © Steve Selvin 2019. Published in 2019 by Oxford University Press. DOI: 10.1093/oso/9780198833444.001.0001

more outlier/extreme values, which is especially important in analysis of small samples of data. Occasionally, a logarithmic scale reduces sensitivity to the influence of outlier/extreme values providing a more accurate analysis. Other kinds of approaches require assumptions about the population sampled. As always, when assumptions are at least approximately correct these techniques can be useful. However, such statistical assumptions are often difficult to rigorously justify.

By far the most famous outlier/extreme value occurred during the US 41st presidential election of the year 2000. Late election night returns showed candidates Albert Gore (Democrat) and George W. Bush (Republican) essentially tied based on current tabulation of the nationwide electoral college vote. The count from the state of Florida was delayed and would decide the next president of the United States.

The Florida county Palm Beach had revised its ballot to make voting simpler and easier to use, particularly for the elderly. The new ballot, called the "butterfly ballot," had the opposite effect. The newly designed ballot caused voters to mistakenly vote for Patrick Buchanan (a third party candidate) when their intention was to vote for Albert Gore. It was estimated that close to 2800 Buchanan votes were such errors making the Bush/Gore vote ratio the exceptionally large value of 1.7. Buchanan himself said, "there must be a mistake." The Democratic party sued. The Florida supreme court ruled in its favor and allowed a recount that would elect Albert Gore. The Republican party appealed the Florida court decision to the Republican dominated US supreme court and won. George W. Bush became the president of the United States.

The Florida statewide final Bush/Gore vote ratio was:

$$Bush/Gore = ratio = \frac{2,912,790}{2,912,253} = 1.09.$$

The US supreme court rejected a recount of disputed votes giving Bush a 537 vote statewide lead from close to 6 million votes (0.1%).

Most descriptions of the Palm Beach ballot and the 2000 presidential election contain the word "contentious" which is certainly an understatement. The entire 36-day affair, involving undoubtedly one of the most critical outlier/extreme value decisions, is described in detail in a book entitled "Too Close to Call."

One last note: Home runs?

For the last several years total number of major league baseball home runs declined each season. Then, in the 2015 season, the trend dramatically reversed. The number of home runs hit in major league baseball games increased by 17%. That is, a total of 723 more home runs hit than the previous season. No evidence emerged to declare the increase as an outlier or an extreme value. Like many outlier/extreme values, lots of speculation but no concrete answers.

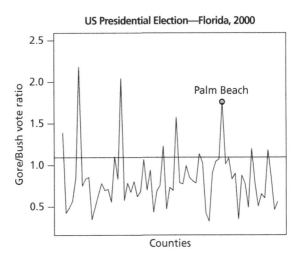

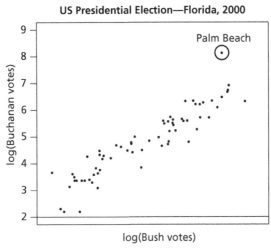

Reference: *Too Close to Call*, by Jeffrey Toobin, Random House, 2001.

7

The role of summary statistics
—brief description

Early 20th century statisticians began to focus on analyses of estimated summary statistics to describe and explore collected data. Mean values, for example, were used to summarize large cumbersome data sets using a few summary values to make succinct and rigorous comparisons among groups. The comparison of two mean values, for example, denoted $\bar{x}_1 - \bar{x}_2$, replaced comparing distributions of observations from two sometimes extremely large sets of data.

Data consisting of large numbers of pairs of observations were similarly summarized by a straight line or a specific curve, leading again to simple and useful summary descriptions enhancing understanding of pairwise x/y-data (called bivariate observations).

Data sets consisting of three variables were summarized by an estimated plane. For example, measures of maternal height and weight are important elements in the description of infant birth weight. Behavior of an estimated plane, similar to the mean value and estimated line, can be used to summarize and describe the joint relationship, producing an analysis of sets of three variables (called multivariate or trivariate observations). Specifically, the slope of one side of the estimated plane could describe the influence of maternal height and the slope of the other side the influence of maternal weight.

The creation of summary relationships for more complex analyses requires more extensive theory and is mathematically more difficult, but the principle of using summary relationships (sometimes called statistical models) as a data analytic tool is central to statistical methods.

The mean value is a familiar and naturally calculated summary of a set of observations. A straight line similarly provides a familiar and easily estimated summary of pairs of observations. That is, a line, like the mean value, succinctly summarizes a set of n paired x/y-observations often providing a clear and simple picture of the relationship between pairs of variables. Several techniques exist to estimate such summary lines.

The Joy of Statistics: A Treasury of Elementary Statistical Tools and their Applications. Steve Selvin. © Steve Selvin 2019. Published in 2019 by Oxford University Press. DOI: 10.1093/oso/9780198833444.001.0001

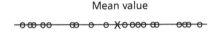

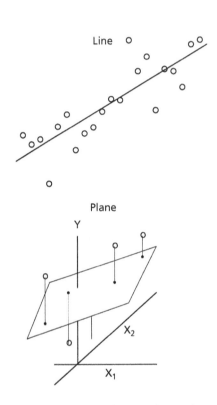

A method requiring estimation of six median values painlessly produces a robust estimate of a straight line. Robust in a statistical context means resilient to influences of extremely small and large values that potentially have considerable and divisive influences on analyses based on summary statistics. As noted, the median is simply the middle value of a distribution of ordered observations regardless of their magnitude.

The estimation of a summary line from paired observations

Three relevant median values calculated from data generated x-values are:

The median value of the left-half of the x-values (denoted L_x).
The median value of all n observed x-values (denoted M_x).

The median value of the right-half of the x-values (denoted R_x).
Three median values are also calculated from the observed y-values (denoted L_y, M_y, and R_y). The estimated line summarizing n pairs of values (x_i, y_i), denoted $y_i = A + B_i$, has intercept and slope represented by A and B.

Artificial data ($n = 21$ x/y-observations):

Table 7.1 Pairs of observations (n = 21)

X	1.0	2.0	3.0	4.0	5.0	**6.0**	7.0	8.0	9.0	10.0	**11.0**	12.0
Y	7.6	18.5	20.4	22.0	23.4	**23.9**	25.9	26.1	30.1	31.0	**31.5**	34.7
X	13.0	14.0	15.0	**16.0**	17.0	18.0	19.0	20.0	121.0			
Y	35.1	39.0	44.1	**44.2**	44.8	48.9	52.9	54.3	154.4			

Note: bold = the six median values

Specifically, x-value medians are:

$$L_x = 6.0, M_x = 11.0, \text{ and } R_x = 16.0$$

and y-value medians are:

$$L_y = 23.9, M_y = 31.5, \text{ and } R_y = 44.2.$$

As expected, all six median values are not adversely affected by the intentionally included extreme outlier 21st pair (121.0, 154.4).

The slope of a straight line is the ratio of the difference between any two y-values divided by the difference between the two corresponding x-values. In symbols,

$$\text{slope} = B = \frac{y_i - y_j}{x_i - x_j} \text{ for any } x/y - \text{pairs}$$

Therefore, an estimated slope $\hat{B}$ based on median values is:

$$\hat{B} = \frac{R_y - L_y}{R_x - L_x}.$$

The relationship, again based on the slope of the line denoted B, where:

$$B = \frac{M_y - A}{M_x - 0}$$

yields an estimate of intercept $\hat{A}$ as:

$$\hat{A} = M_y - \hat{B}(M_x).$$

For the example data:

$$\hat{B} = \frac{44.2 - 23.9}{16 - 6} = 2.0 \quad \text{and} \quad \hat{A} = 31.5 - 2.0(11) = 9.5.$$

The robust estimated line $\hat{y} = \hat{A} + Bx$ becomes $\hat{y} = 9.5 + 2.0x$.

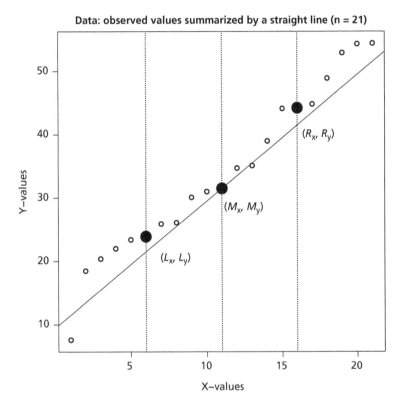

Data: observed values summarized by a straight line (n = 21)

The progression of men's and women's world records in swimming (4 × 100 meters freestyle relay) illustrate the clarity achieved by comparing summary straight lines (Internet data). These estimated lines clearly identify that women's world record times decreased faster than comparable men's times over the years 1920 to 2015. The estimated straight lines reveal men's times decreased by 1.1 minutes over the almost 100-year period while over the same period women's times decreased by 1.8 minutes.

It might be entertaining to calculate the year men's and women's times would be essentially equal making it possible for direct male/female competition. The estimated lines intersect at year 2035, predicting equal world record times. However, this prediction depends on the estimated straight lines continuing to accurately represent decrease in world record times, which is not likely. If women's times continued the same linear decline the winning time by the year 2178 would be reduced to zero. It is important to note, and is also an important general rule, that extrapolating beyond limits of observed data is almost always unreliable and likely misleading. A popular song from the mid-1950's entitled *Que Sera, Sera*, wisely suggested:

> "whatever will be, will be
> the future's not ours to see."

If statistical analyses produced accurate predictions beyond the range of collected data, a fortune could be made betting on horse racing or sports events, or buying stocks, or just about anything.

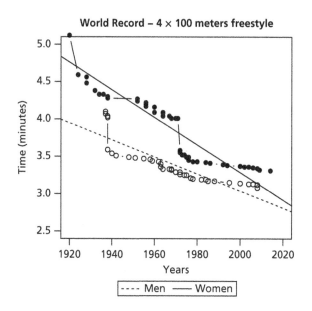

8

Correlation and association
—interpretation

Measures of correlation and association are central to both identifying and summarizing relationships within pairs of observations. Frequently the terms "correlation" and "association" are used interchangeably but statistically they have distinct applications and interpretations.

Table 8.1 Example correlation coefficients (r) from a coronary heart disease study

age	r
age and height	−0.10
age and weight	−0.03
age and cholesterol	0.09

blood pressure	r
diastolic and systolic	0.77
systolic and age	0.17
systolic and weight	0.25
systolic and cholesterol	0.17
diastolic and age	0.14
diastolic and weight	0.30
diastolic and cholesterol	0.13

An unambiguous and fundamental property of a correlation coefficient (traditionally denoted r) is that it only measures the extent a straight line represents the x/y-relationship between x/y-data pairs. The variables x and y are continuous variables, typically measured values such as height, weight, time, and distance. The almost metaphysical definition of a continuous variable is:

a variable is continuous if between any two values there exists a third value.

The Joy of Statistics: A Treasury of Elementary Statistical Tools and their Applications. Steve Selvin. © Steve Selvin 2019. Published in 2019 by Oxford University Press. DOI: 10.1093/oso/9780198833444.001.0001

The statistical expression to estimate a correlation coefficient can be found in any introductory statistics textbook and all statistical computer packages, and may be calculated using Internet sites. Its mathematical origins and statistical properties are not simply described.

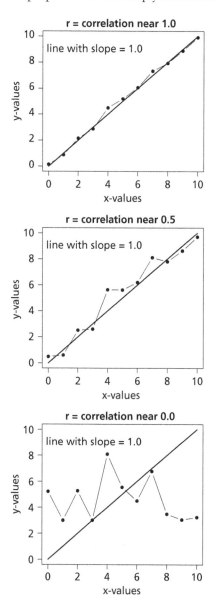

A correlation coefficient equal to one occurs when the x/y-pairs are exactly represented by a straight line. A correlation of zero indicates variables x and y are not linearly related. The word "linear" is important. Situations exist where variables x and y have a small or zero correlation but are related in nonlinear ways. For example, the four corners of a rectangle, where $x = \{1, 4, 4, 1\}$ and $y = \{1, 1, 2, 2\}$, are related but the correlation between values x and y is zero. Only a linear relationship between two variables is characterized by a single correlation coefficient ranging between -1 and $+1$.

The magnitude of a correlation coefficient unequivocally indicates the extent of a linear relationship between two variables but requires care to be usefully interpreted without further information. For example, systolic blood pressure is correlated with body weight ($r = 0.25$). Diastolic blood pressure is similarly correlated with body weight ($r = 0.30$). Without further analysis, it is not possible to separate two possibilities. Either systolic blood pressure is directly correlated with body weight or, because systolic blood pressure is correlated with diastolic blood pressure ($r = 0.77$), it appears to be correlated with body weight ($r = 0.30$). That is, the observed correlation between systolic blood pressure and body weight is due indirectly to its correlation with diastolic blood pressure which is correlated with body weight. To usefully separate and measure influences of correlated variables, such as systolic and diastolic blood pressure, multivariable statistical techniques are necessary.

A spurious correlation

An extreme example:

Correlation coefficients close to $+1$ or -1 that exist by chance or by manipulation are called spurious correlations. An exaggerated case emphasizes the fact that a correlation close to one does not necessarily indicate a meaningful relationship. An entire Internet website is devoted to a large number of examples of these nonsensical spurious correlations.

One example:

Table 8.2 Uranium stored in US nuclear plants (x) and math doctorates awarded (y)

year	1996	1997	1998	1999	2000	2001	2002	2003	2004	2005	2006	2007	2008
x	66.1	65.9	65.8	58.3	54.8	55.6	53.5	45.6	57.7	64.7	77.5	81.2	81.9
y	1122	1123	1177	1083	1050	1010	919	993	1076	1205	1325	1393	1399

correlation coefficient $= r = 0.952$

Association

A relationship between two discrete variables is called an association. Discrete variables can be either counts or categorical classifications such as months or years or educational attainment or ethnicity. Association between such variables is frequently important and, similar to a correlation coefficient, not easily interpreted. Again, underlying and unmeasured variables can substantially influence observed pairwise associations.

Artificial data describing an association between two variables A and B occurring between husband and wife at three different ages illustrate:

Table 8.3 Artificial data

		husband		
		A	$\bar{A}$	*total*
wife	B	66	54	120
	$\bar{B}$	54	126	180
	total	120	180	300

The probability B occurs when A has occurred is $P(B|A) = 66/120 = 0.55$ (first column in the table). The probability B occurs is $P(B) = 120/300 = 0.40$, indicating an A/B-association. In other words, occurrence of A increases the likelihood of occurrence of B (0.40 versus 0.55). Thus, variables A and B are associated, positively associated.

Three artificial tables illustrate the difficulty of interpreting pairwise measures of association.

Table 8.4 Age 30–50 years

		husband		
		A	$\bar{A}$	*total*
wife	B	1	9	10
	$\bar{B}$	9	81	90
	total	10	90	100

The probability B occurs when A has occurred is $1/10 = 0.10$ and the probability of B is $10/100 = 0.10$ indicating exactly no A/B-association, variables A and B are independent.

Table 8.5 Age 50–70 years

		husband		
		A	$\bar{A}$	total
wife	B	16	24	40
	$\bar{B}$	24	36	60
	total	40	60	100

The probability B occurs when A has occurred is $16/40 = 0.40$ and the probability of B is $40/100 = 0.40$, indicating exactly no A/B-association, variables A and B are independent.

Table 8.6 Age 70–90 years

		husband		
		A	$\bar{A}$	total
wife	B	49	21	70
	$\bar{B}$	21	9	30
	total	70	30	100

The probability B occurs when A has occurred is $49/70 = 0.70$ and the probability of B is $70/100 = 0.70$, indicating exactly no A/B-association, variables A and B are independent.

The three example tables demonstrate a third variable can be the source of an apparent A/B-association. Combining Tables 1, 2, and 3, each exhibiting exactly no A/B-association, creates a table with a strong association, the first husband/wife table (Table 0). Specifically, in this exaggerated example, the variable of age is completely the source of observed A and B husband/wife association (Table 0). Increasing frequency of variables A and B with increasing age causes a corresponding increase in the joint occurrence of variables A and B, producing an apparent A/B-association. Often, a number of variables contribute to pairwise associations and, like a correlation coefficient, more extensive analysis is typically necessary to accurately identify and separate these influences to produce a clear sense of an association.

To summarize:

A phrase sometimes used to describe properties of a mathematical proof applies to the relationship between correlation and cause—it is said,

the conditions are necessary but not sufficient.

Less formally, a high correlation rarely indicates a direct causal linear relationship. That is, a high correlation is a necessary condition but rarely sufficient because of the always present possibility of influences from additional unmeasured correlated variables.

Two possibilities—a correlated A to B

The story of three correlation coefficients—XZ, YZ, and XY

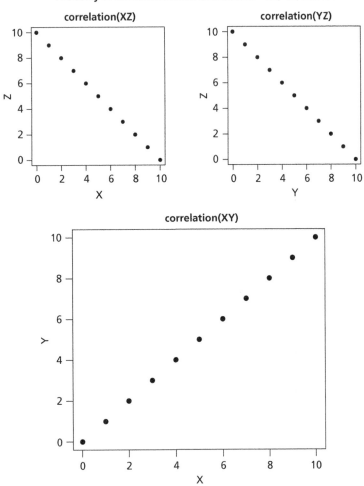

9

Proportional reduction in error —a measure of association

The correlation coefficient is the famous measure of a linear relationship between pairs of observations but is not useful for assessing association between discrete variables used to create a table of counts. Statisticians Goodman and Kruskal produced a simple and intuitive measure of association specifically for this task.

A 1960 study from western New York state indicated socioeconomic status (denoted *ses*) was associated with occurrence of lung cancer. Increasing socioeconomic status appeared to be associated with increasing lung cancer rates. Smoking/*ses* data were collected to determine if cigarette smoking was a factor in this observation.

Table 9.1 Smoking exposure and levels of socioeconomic status ($n = 2422$)

	ses_1	ses_2	ses_3	ses_4	ses_5	total
smokers	78	124	149	221	343	915
nonsmokers	102	153	220	252	481	1208
past smokers	103	81	32	48	35	299
total	283	358	401	521	859	2422

From the table, key statistics are the maximum values in each row. The notation is: M_0 is the maximum count in the total row and M_i-values are maximum counts in each of the other rows. For the smoking/ses data:

$$M_0 = 859 \text{ and } M_1 = 343, M_2 = 481, M_3 = 103.$$

Then, a measure of association between row and column variables, based on the Goodman/Kruskal statistical summary called *proportional reduction in error*, is:

$$\hat{R} = \frac{\sum M_i - M_0}{n - M_0} = \frac{343 + 481 + 103 - 859}{2422 - 859} = \frac{68}{1563} = 0.044.$$

The Joy of Statistics: A Treasury of Elementary Statistical Tools and their Applications. Steve Selvin. © Steve Selvin 2019. Published in 2019 by Oxford University Press. DOI: 10.1093/oso/9780198833444.001.0001

Proportional reduction in error—how it works

An often used statistical strategy is to calculate a summary value under one set of conditions and then calculate the same summary value under a second set of conditions. Comparison measures the extent of difference between conditions.

The Goodman/Kruskal proportional reduction in error measure (denoted $\hat{R}$) addresses the question: Does level of one variable (smoking status) influence another variable (socioeconomic status)? The comparison is between excluding smoking from consideration (M_0) to accounting for smoking exposure (M_1, M_2, and M_3). The smoking/ses data yields only a slight indication of such an association ($\hat{R} = 0.044$).

Two measures based on maximum values from a table of counts provide a statistical comparison. One choice is to measure increase associated with the "best guess" of social class based on each row (for example, for smokers, $343/915 = 0.375$). Another choice is based on error reduction using the complementary probability of the "best guess" (for example, for smokers $1-343/915 = 0.625$). The difference in reduction of error produces a compact and simple analytic expression measuring the degree of association between smoking and socioeconomic status, again denoted $\hat{R}$. Specifically, conditions influencing the extent of error are compared when influence of smoking is ignored to the conditions influencing the extent of error when influence of smoking is considered.

The Goodman/Kruskal approach starts with a probability (denoted P) that reflects the extent of error in "guessing" social class when smoking exposure is ignored. Thus, $P = 1-859/2422 = 0.645$ where 859 is the largest count among study participants ignoring smoking status (maximum count in the total row, namely M_0).

Parallel error probabilities calculated using the potential influence, if any, from smoking status are:

$P_1 = 1-343/915 = 0.625$ where 343 is the largest count of subjects who smoke.

$P_2 = 1-481/1208 = 0.602$ where 481 is the largest count of subjects who do not smoke.

$P_3 = 1-103/299 = 0.656$ where 103 is the largest count of subjects who are past smokers.

Then, comparing the probability (P) based on the maximum value ignoring any influence of smoking (last row) to the row probabilities based on the maximum values including the influence of smoking (P_i) measures reduction in error. The probabilities that account for smoking status are combined into a single estimate using a weighted average. Specifically, using the row totals as weights, this probability is:

$$\overline{P} = \frac{1}{2422}\left[915(0.625) + 1208(0.602) + 299(0.656)\right] = 0.617.$$

The comparison P (smoking status ignored) = 0.645 to $\overline{P}$ (smoking status considered) = 0.617 produces proportional reduction in error:

$$\hat{R} = \frac{P - \overline{P}}{P} = \frac{0.645 - 0.617}{0.645} = 0.044.$$

A bit of algebra yields the same estimate $\hat{R}$ calculated from a simple and direct expression based exclusively on the maximum counts from each row in a table, namely:

$$\hat{R} = \frac{\sum M_i - M_0}{n - M_0}.$$

For the example smoking/ses table, again:

$$\hat{R} = \frac{[343 + 481 + 103] - 859}{2422 - 859} = \frac{927 - 859}{2422 - 859} = \frac{68}{1563} = 0.044.$$

Another example:

Table 9.2 Case/control trial of vitamin C as a cold prevention ($n = 237$)

colds	0	1–3	4+	total
vitamin C	46	90	22	158
placebo	23	45	11	79
total	69	135	33	237

Where $M_1 = 90$, $M_2 = 45$, and $M_0 = 135$, then proportional reduction of error is:

$$\hat{R} = \frac{M_1 + M_2 - M_0}{n - M_0} = \frac{90 + 45 - 135}{237 - 135} = 0.$$

These case/control clinical trial data are actual values from a Canadian study of the influence of vitamin C on the frequency of winter colds. Furthermore, it is certainly exceptional that the data naturally resulted in perfect independence between case/control status and number of reported colds. The proportional reduction in error is exactly $\hat{R} = 0$.

When row and column variables are exactly independent, all row maximum counts occur in the same column. Then, the sum of these maximum values also is the maximum of the column sums producing an $\hat{R}$ value of zero. In symbols, from the Canadian data, then $M_1 + M_2 = M_0$ and $M_1 + M_2 - M_0 = 90 + 45 - 135 = 0$.

One last example:

Table 9.3 Mother/daughter—educational attainment ($n = 759$)

attainment	less than HS	HS	more than HS	total
less than HS	84	100	67	251
HS	142	106	77	325
more than HS	43	48	92	183
total	269	254	236	759

Note: HS = high school education only

Summary maximum counts: $M_0 = 269$ and $M_1 = 100$, $M_2 = 142$, $M_3 = 92$, then:

$$\hat{R} = \frac{M_1 + M_2 - M_0}{n - M_0} = \frac{100 + 142 + 92 - 269}{759 - 269} = \frac{65}{490} = 0.133.$$

Reference: *The Analysis of Cross-Classified Data Having Ordered Categories*, by Leo Goodman, Harvard University Press, 1984

10

Quick tests—four examples

A fundamental statistical question:

Do sampled data reveal evidence of a systematic influence?

Four quick tests address this question without assumptions about properties of the population sampled or tables of probabilities or mathematical computations, and without involving extensive statistical theory or use of computer software. Occasionally these tests are called "pocket tests." Like a pocket watch or a pocket comb, these statistical tools are simple, handy, useful, and easily applied. John Tukey's quick test has all these properties.

John Tukey (b. 1915) is one the most important statisticians of the 20th century. He is well known for his boxplot and stem–leaf plot as well as a long list of valuable statistical methods that are today taken for granted. Equally amazing is his list of national and international activities and awards. For example, he was a key participant in negotiations that established the Russian/US nuclear arms treaty.

Tukey's quick test

Tukey's quick test is designed to compare two independent samples of data to explore the question: do data sampled from two sources reveal evidence of systematically differing distributions? Sometimes called a "two-sample comparison." The principle underlying this quick test is simple. If observed values from two sources have the same distribution, the two sets of sampled values are likely similar. Alternatively, when values come from different distributions, differences of extreme values will likely identify differences between two samples of observed values. Specifically, when the samples differ, one sample likely has more small values and the other sample likely has more large values.

Tukey suggested a simple statistical comparison. Starting with the sample containing the smallest value, count the number of values less than the smallest value in the other sample (denoted t_A). Similarly,

The Joy of Statistics: A Treasury of Elementary Statistical Tools and their Applications. Steve Selvin. © Steve Selvin 2019. Published in 2019 by Oxford University Press. DOI: 10.1093/oso/9780198833444.001.0001

count the number of values in the second sample greater than the largest value in the first sample (denoted t_B). The sum $T = t_A + t_B$ measures the degree of overlap/separation between two sets of sampled values. Large values of T likely identify a nonrandom difference.

Artificial data illustrate:

A	0.2	1.5	2.1	4.4	4.0	6.7	3.0	4.8	5.3	4.2
B	6.6	3.4	5.3	4.1	3.7	5.8	6.3	3.1	7.9	7.8

The same data sorted:

A	[0.2	1.5	2.1	3.0]	4.0	4.2	4.4	4.8	5.3	6.7
B	3.1	3.4	3.7	4.1	5.3	5.8	6.3	6.6	[7.8	7.9]

Note: bold values identify non-overlapping values.

The value $t_A = 4$ (namely, A-values 0.2, 1.5, 2.1, and 3.0 are less than the smallest B-value 3.1). The value $t_B = 2$ (namely, B-values 7.8 and 7.9 are greater than largest A-value 6.7). Tukey's summary statistic yields a measure of difference $T = t_A + t_B = 4 + 2 = 6$.

Assessment:

A test statistic T equal to or greater than six occurs with probability of approximately 0.05 or less when two sampled distributions differ by chance alone. That is, a value of T with a probability of occurrence of less than 0.05 indicates the T-measured shift is unlikely due entirely to random variation, providing evidence of a systematic difference ($T \geq 6$).

The larger the value T, the stronger the evidence.

Technical issues:

1. Tukey's quick test requires no tied values occur.
2. Sample A contains the smallest observation.
3. Sample B contains the largest observation.
4. The two samples are approximately the same size.
5. The sampled observations are independent.

Statistical power is defined as the probability of identifying a difference when a difference exists. Statistician Churchill Eisenhart defined *practical statistical power* as statistical power which also includes the likelihood the statistical method is used. In other words, an easily applied and clearly understood quick test method often compensates for loss of statistical detail and power of more extensive complicated methods.

Correlation—a quick test version

Unpublished data from the best ten ($n = 10$) major league baseball teams (2014 season) consisting of games won and total wages paid to players (in millions of dollars) provide an illustration of a quick test analysis based on an estimated correlation coefficient (denoted r).

Table 10.1 Baseball teams, wins, and wages

team	wins	wages
Los Angeles	92	272
New York	87	219
Boston	78	187
Detroit	74	173
San Francisco	84	172
Washington	83	164
St. Louis	85	151
Texas	88	142
Philadelphia	63	135
Toronto	93	222

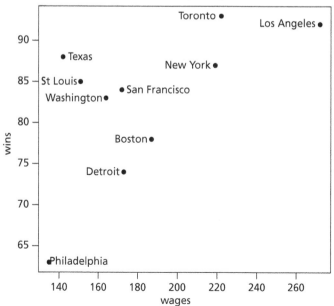

Wins—top ten major league teams

A graphical display of the relationship between wins and wages is a useful description but lacks a quantitative assessment of possible influences from random variation.

The estimated correlation coefficient $r = 0.603$ indicates a potential relationship between games won and wages paid but also does not formally account for possible random influences on the win/wages relationship.

Like Tukey's quick test, a correlation quick test is a quantitative assessment using a simple comparison between an observed correlation coefficient (denoted r) and a specific statistical bound (denoted r_0).

When the value of an estimated correlation coefficient based on n pairs of observations is due entirely to random variation, the probability the estimated value r exceeds $r_0 = 1.645/\sqrt{n-2}$ is approximately 0.05 or less. Like the pattern of most quick tests, an estimated correlation coefficient greater than r_0 produces the inference that the observed correlation is unlikely due entirely to random variation. Thus, a quick test again becomes a comparison of the estimate r calculated from data to a specially created bound r_0. The observed correlation coefficient of $r = 0.603$ from the wages/wins data exceeds $r_0 = 1.645/\sqrt{n-2} = 1.645/\sqrt{8} = 0.582 (r > r_0)$ and is, therefore, unlikely to have occurred by chance alone. More formally, the probability the observed correlation coefficient 0.603 resulted from purely a chance relationship between wins and wages is slightly less than 0.05.

Sign test—another quick test

A subset of data ($n = 20$) from an extensive before/after intervention trial of an experimental strategy to reduce cigarette smoking among extremely addicted men illustrates. Smoking exposure was measured before and after an introvention program by change in levels of thiocyanate in participant's blood. The only source of blood levels of thiocyanate is cigarette smoking.

The sign test, as the name suggests, starts by replacing observed numeric differences with a plus or minus sign as a non-numeric measure of within-pair difference. If only random differences occur within all before/after pairs, the number of positive and negative signs would differ randomly. However, if the introvention program reduces smoking exposure, the count of positive signs (reduced thiocyanate levels) would likely exceed $n/2$.

Table 10.2 Before/after introvention data—thiocyanate levels (n = 20)

pairs	1	2	3	4	5	6	7	8	9	10	11	12	13	14	15	16	17	18	19	20
before	131	154	45	89	38	104	127	129	148	181	122	120	66	122	115	113	163	67	156	191
after	111	13	16	35	23	115	87	120	122	228	118	121	20	153	139	103	124	37	163	246
difference	20	141	29	54	15	-11	40	9	26	-47	4	-1	46	-31	-24	10	39	30	-7	-55
sign	+	+	+	+	+	-	+	+	+	-	+	-	+	-	-	+	+	+	-	-

Note: + = reduction of thiocyanate level

The count of positive pairs, when no association exists, exceeds the value $c_0 = n/2 + 0.823\sqrt{n}$ with an approximate probability of 0.05 or less. A quick test again becomes a comparison of the count of signs (denoted c) to a calculated bound (denoted c_0). The value $c_0 = n/2 + 0.823\sqrt{n} = 10 + 0.823\sqrt{20} = 13.68$ for $n = 20$ observations. For the introvention trial data, the observed count of positive signs is $c = 13$. Therefore, the quick test outcome of 13 positive signs compared to bound 13.68 (c versus c_0) occurred by chance with a probability slightly more than 0.05 ($c < c_0$). Thus, the sample data fail to produce substantial evidence of a systematic influence from the experimental introvention strategy. In other words, chance remains a plausible explanation for the observed increase in the number of positive differences.

Quick test—a two sample visual comparison

A cousin of Tukey's two-sample quick test is a graphical comparison. The comparison starts with ordering observations in each of two samples of equal size from smallest to largest producing a series of x/y-pairs.

A plot of pairs creates a straight line with slope one and intercept zero when all pairs consist of identical values. A plot of observed x/y-pairs displays the extent of deviation from this theoretical line producing a simple and direct graphical display of the difference between two compared data sets.

Body mass index (*bmi*) measurements from 27 white (x) and Hispanic women (y) illustrate this two-sample comparison. A plot of the 27 x/y-pairs produces no obvious visual evidence of important differences between the two samples.

Table 10.3 White/Hispanic women—bmi measurements

white	17.9	18.6	19.2	19.2	19.7	20.5	21.1	21.3	21.7
Hispanic	17.8	18.7	19.3	20.0	20.1	20.1	20.7	21.4	21.6
white	21.8	22.4	22.9	22.9	23.0	23.0	23.1	23.6	23.8
Hispanic	21.8	22.0	22.0	22.1	22.6	22.9	23.1	23.1	23.3
white	24.5	24.7	24.8	26.5	27.2	28.2	29.1	29.3	30.2
Hispanic	23.6	24.0	25.0	25.4	25.6	25.9	26.7	27.4	31.3

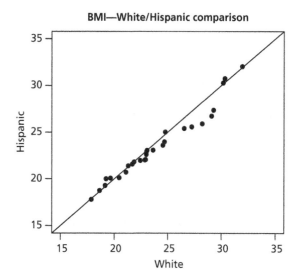

BMI—White/Hispanic comparison

The four quick tests, like all quick tests, sacrifice precision and are generally approximate. However, as the names suggest, they often provide a simple and direct answer to a statistical question. That is, quick tests often have high levels of practical statistical power. As usual, these four quick tests become more precise as the sample size increases.

Reference: Unpublished before/after thiocyanate data from the 1982 clinical trial called Mrfit (Multi risk factor introvention trial) of coronary heart disease risk factors.

11

Confounding—African-American and white infant mortality

California state vital records include number of live births and deaths of African-American and white infants. A perinatal mortality rate is:

$$perinatal\ mortality\ rate = \frac{infant\ deaths\ before\ 28\ days + fetal\ deaths}{live\ births + fetal\ deaths}.$$

Data: from California birth and death certificates of black and white infants:

Table 11.1 Perinatal mortality rates per 1000 live births, California (2008)

white infants			black infants		
deaths	births	rate	deaths	births	rate
2897	404,023	7.17	520	44,356	11.72

rate ratio = black/white = $r = 11.72/7.17 = 1.635$

Black/white comparison of perinatal mortality rates is not as simple as it looks. The compared rates are crude mortality rates. The word "crude" does not mean raw or unrefined. The statistical use of the word crude simply means unadjusted for potentially obscuring and unaccounted influences. Such a third variable makes comparison of crude black and white perinatal mortality rates a misleading measure of risk, a very misleading measure.

The word "confound" is a verb meaning causes surprise or confusion. The word "confounder" is a noun used in a statistical/epidemiological context to refer to a variable which complicates interpretation of relationships among the variables under study. Specifically, direct comparison of perinatal mortality rates between newborn black and white infants is biased because birth weights of newborn black infants are typically less

The Joy of Statistics: A Treasury of Elementary Statistical Tools and their Applications. Steve Selvin. © Steve Selvin 2019. Published in 2019 by Oxford University Press. DOI: 10.1093/oso/9780198833444.001.0001

than newborn white infants and mortality risk increases as birth weight decreases. The difference in black/white infant mean birth weights is about 250 grams (approximately a half-pound). Thus, direct comparison of black/white infant mortality requires a statistical approach that accounts for excess mortality of black infants caused by their naturally lower birth weights, which increases their observed mortality rates. Thus, low birth weight black infants cause a considerable confounding influence.

The simplest and most direct approach is to classify infants into a series of narrow intervals based on birth weight. The Californian mortality data classified into 100 gram intervals produces 35 interval-specific black/white rate ratios. Within each interval the difference between black and white infant birth weights is small and negligible. Therefore, black/white birth weight differences have almost no confounding influence on each of these 35 estimated rate ratios (denoted r_i).

Results:

1. Within 35 birth weight intervals, 27 perinatal black rates are less than white rates indicating generally lower mortality. The within-interval comparison accounts for differing birth weights. Specifically, from 35 rate ratios, 27 are less than one (77%).

2. A plot of black and white rates shows for most birth weights black perinatal mortality rates are lower than white rates and, to repeat, the interval-specific comparisons are minimally influenced by differing black/white birth weights. Ratios greater than one generally occur among the high birth weight black infants (>3500 g) where only very few observations occur and are likely due to random variation.

3. The plot of the logarithms of perinatal rates reveals two important features. The black/white mortality curves are approximately parallel indicating black/white mortality ratios are close to constant for all birth weights. Also, black mortality log-rates reveal the substantial increased variability beyond birth weights of 3500 grams due to extremely small numbers of high birth weight black infants (3.5%).

The mean perinatal mortality black/white rate ratio is 0.754, accounting for differences in birth weight. Specifically,

$$mean\ log-rate = \overline{log[R]} = \frac{1}{35}\sum log[r_i] = \frac{1}{35}(-9.853) = -0.282.$$

The estimated mean ratio becomes $\overline{R} = e^{-0.282} = 0.754$. To repeat, this summary ratio $\overline{R}$ is the mean of 35 separate and approximately equal interval-specific estimates that consist of infants with essentially the same birth weights.

The remarkable reversal from the unadjusted black/white comparison (0.754 adjusted versus 1.635 unadjusted) is due to accounting for the exceptionally strong confounding influence of black infant birth weights.

Table 11.2 Data: black and white deaths, births, perinatal mortality rates/1000 births (California—2008) and rate ratios (black/white) for 35 birth weight categories (100 gram intervals)

birth weights	white infants			black infants			ratios
	deaths	births	rates	deaths	births	rates	
≤ 900	173	322	537.27	65	131	496.18	0.924
901–1000	148	337	439.17	40	122	327.87	0.747
1001–1100	134	398	336.68	30	131	229.01	0.680
1101–1200	106	381	278.22	29	137	211.68	0.761
1201–1300	103	444	231.98	21	143	146.85	0.633
1301–1400	86	427	201.41	19	143	132.87	0.660
1401–1500	108	597	180.90	19	165	115.15	0.637
1501–1600	85	560	151.79	20	167	119.76	0.789
1601–1700	84	682	123.17	24	219	109.59	0.890
1701–1800	86	722	119.11	12	194	61.86	0.519
1801–1900	100	935	106.95	26	298	87.25	0.816
1901–2000	81	978	82.82	15	299	50.17	0.606
2001–2100	74	1589	46.57	21	420	50.00	1.074
2101–2200	87	1714	50.76	10	453	22.08	0.435
2201–2300	82	2322	35.31	14	603	23.22	0.657
2301–2400	80	2885	27.73	12	763	15.73	0.567
2401–2500	80	4149	19.28	13	977	13.31	0.690

Reference: State of California Vital Records, 2008 by ethnic group.

African-American and white infant mortality rates by birth weight

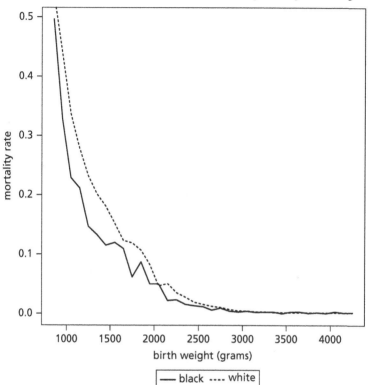

African-American and white infant mortality log-rates by birth weight

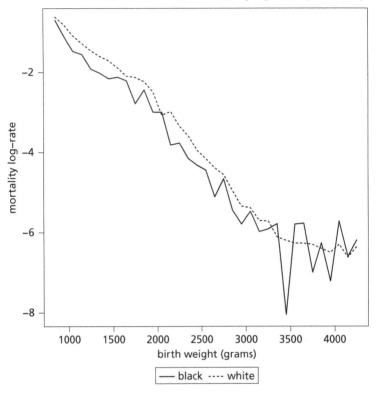

12

Odds—a sometimes measure of likelihood

Odds are a ratio measure of the likelihood of a specific outcome. Odds "for" measure the likelihood an outcome occurs. Odds "against" measure the likelihood an outcome does not occur. The odds* 1:5 are the probability $1/6 = 0.167$ divided by the complementary probability $= (1–1/6) = 5/6 = 0.833$.

In general, odds are defined as a probability divided by one minus the same probability. In symbols, the ratio is:

$$odds = \frac{p}{1-p}$$

where p represents a probability of a specific outcome.

Odds in the context of gambling are clearly illustrated by horse race betting. A typical horse racing betting form (also posted at the track at race time with current odds) is:

Table 12.1 Typical racing form—Golden Gate Fields (2008)

pp**	horse	jockey	wt	odds	"probability"
2	Trible Headress	A. Gorrez	122	5-2	0.29
4	California Wildcat	R. Gonzales	120	4-1	0.20
1	Fastasme	R. Baze	120	3-1	0.25
7	Madame Cluet	J. Couton	126	5-1	0.17
6	Cat Over Fence	F. Alvado	118	6-1	0.14
3	Hot Jem	F. Crispin	116	8-1	0.11
8	Port Elizabeth	K. Frey	121	15-1	0.06
5	Anget Elia	W. Armongeogi	120	20-1	0.05

** pp = post position
Note: the sum of the "probabilities" is 1.27.

* = Odds are a single quantity that is traditionally treated as a plural noun.

The Joy of Statistics: A Treasury of Elementary Statistical Tools and their Applications. Steve Selvin. © Steve Selvin 2019. Published in 2019 by Oxford University Press. DOI: 10.1093/oso/9780198833444.001.0001

If these "probabilities" added to one, horse race betting would be a fair game. The amount of money bet would exactly equal the amount of money paid to winners. However, horse racing odds when transformed to "probabilities" always add to more than one, indicating from money bet it is necessary to sustain the race track, award the winning horse owner, and, of most importance, make a profit. In short, horse race gambling is not a fair game. In addition, odds are not determined by considerations such as the last performance of the horse or the reputation of the jockey or number of previous wins. The odds are entirely determined by the amount of money bet on each horse calculated to guarantee a profit from every race. The example shows this amount of the total bet is about 27%.

Odds ratio

An odds ratio, as the name suggests, is a ratio of two sets of odds. If the odds are 4:1 for outcome A and 2:1 for outcome B, a summary ratio from these odds is:

$$odds\ ratio = \frac{4/1}{2/1} = 2,$$

creating a comparison of likelihoods of two outcomes A and B in terms of a ratio. Two useful relationships between odds and a probability (p) are:

$$p = \frac{odds}{1 + odds} \quad \text{and} \quad odds = \frac{p}{1 - p}.$$

Commonly, in a statistical context, from a 2×2 table displaying the frequencies of outcomes of joint occurrence of events A and B:

Table 12.2 Notation for two independent events—denoted A and B

	outcome A	outcome $\bar{A}$
outcome B	a	b
outcome $\bar{B}$	c	d

$$\textit{estimated odds ratio} = \frac{a/c}{b/d} = \frac{a/b}{c/d} = \frac{ad}{bc}---\text{typically denoted } \hat{or}.$$

When $p = P(A)$, $P = P(B)$ and outcomes A and B are independent, a 2×2 table of joint probabilities is

Table 12.3 Notation for two independent events—denoted A and B

	outcome B	outcome $\overline{B}$
outcome A	$a = pP$	$b = p(1-P)$
outcome $\overline{A}$	$c = (1-p)P$	$d = (1-p)(1-P)$

and the odds ratio becomes:

$$or = \frac{ad}{bc} = \frac{pP(1-p)(1-P)}{p(1-P)(1-p)P} = 1.0.$$

A bit more:

The odds ratio is a popular numeric ratio measure often used in epidemiologic and genetic analyses. Like the comparison of two mean values $(\overline{x}_1 - \overline{x}_2)$ or two proportions, $(p_1 - p_2)$, an odds ratio is also a summary comparison of data from two sources measured in terms of a single ratio value.

13

Odds ratio—a measure of risk?

Consider data from a study of female military veterans comparing breast cancer incidence between women who served in Vietnam ($n_1 = 3392$) to women who did not ($n_0 = 3038$). The odds ratio reflecting association between breast cancer and kind of military service is $\hat{or} = 1.219$ and the relative risk ratio is $\hat{rr} = 1.208$.

Table 13.1 Data: breast cancer (D) diagnosed among military women who served in Vietnam (E) and who did not serve in Vietnam during the war years (1965 to 1973)

	Breast cancer status		
	D	$\bar{D}$	total
exposed E	$a = 170$	$b = 3222$	3392
unexposed $\bar{E}$	$c = 126$	$d = 2912$	3038
total	296	6134	6430

A relative risk ratio (denoted rr), as the name suggests, measures risk calculated from a 2×2 table. Odds ratio (denoted or) measures association also often calculated from a 2×2 table. Relative risk is the probability of disease among exposed individuals divided by the probability of disease among unexposed individuals. In symbols,

$$\text{relative risk ratio} = rr = \frac{P(\text{probability of disease among exposed individuals})}{P(\text{probability of disease among unexposed individuals})}.$$

The odds ratio is the odds of disease among exposed individuals divided by the odds of disease among unexposed individuals. In symbols,

$$\text{odds ratio} = or = \frac{P(\text{odds of disease among exposed individuals})}{P(\text{odds of disease among unexposed individuals})}.$$

The Joy of Statistics: A Treasury of Elementary Statistical Tools and their Applications. Steve Selvin. © Steve Selvin 2019. Published in 2019 by Oxford University Press. DOI: 10.1093/oso/9780198833444.001.0001

Relative risk and odds ratios estimated from a 2×2 table of counts are:

$$\text{estimated relative risk ratio} = \widehat{rr} = \frac{a/(a+b)}{c/(c+d)}$$

and

$$\text{estimated odds ratio} = \widehat{or} = \frac{a/b}{c/d} = \frac{a/c}{b/d} = \frac{ad}{bc}.$$

When an outcome is rare, the estimated odds ratio becomes approximately equal to the estimated relative risk ratio. Only then can an odds ratio be used as an approximate measure of risk. When these two values calculated from the same data are close to equal, naturally, the odds ratio and relative risk ratio have close to the same properties and similar interpretations.

Comparison of odds ratio and relative risk ratio calculated from Vietnam data illustrates. Because breast cancer is a rare disease, the odds ratio can be treated as an approximate measure of risk.

The odds ratio and relative risk ratio statistics become similar when a, from a 2×2 table, is much less than b ($a << b$) making $a + b$ approximately equal to b and, from the same table, c is much less than d ($c << d$) making $c + d$ approximately equal to d. Then,

$$\widehat{rr} = \frac{a/(a+b)}{c/(c+d)} \approx \frac{a/b}{c/d} = \widehat{or}.$$

From the breast cancer data, since $a + b \approx b$ or $3392 \approx 3222$ ($a = 170$) and $c + d \approx d$ or $3038 \approx 2912$ ($b = 126$), then $\widehat{rr} = \dfrac{170/3392}{126/3038} = 1.208$ approximately equals $\widehat{or} = \dfrac{170/3222}{126/2912} = 1.219$. In symbols, $\widehat{rr} \approx \widehat{or}$.

Nine 2×2 tables illustrate accuracy of an odds ratio as an approximation to a relative risk ratio for increasing numbers of breast cancer cases ($a + c$).

Table 13.2 Odds ratios and relative risk ratios applied to the same table

a	b	c	d	a+c	$\widehat{or}$	$\widehat{rr}$
85	3307	63	2975	148	1.214	1.208
170	3222	126	2912	296	1.219	1.208*
255	3137	189	2849	444	1.225	1.208
340	3052	252	2786	592	1.232	1.208
425	2967	315	2723	740	1.238	1.208
510	2882	378	2660	888	1.245	1.208
850	2542	630	2408	1480	1.278	1.208
1190	2202	882	2156	2072	1.321	1.208
1700	1692	1260	1778	2960	1.418	1.208

* = observed Vietnam data (row 2)

The effects of race and sex on physicians' recommendations for cardiac catherization

METHODS (from the published paper—12 authors):

We developed a computerized survey instrument to assess physician recommendations for managing chest pain. Actors portraying patients with particular characteristics in scripted interviews about their symptoms were recorded. A total of 720 physicians participated in the survey. Each participating physician viewed the recorded interview and was given additional data about the hypothetical patient and asked to make recommendations for cardiac care.

RESULTS (from the abstract of the paper):

Analysis identified adjusted odds ratios for women (odds ratio of $\widehat{or} = 0.60$: confidence interval 0.4 to 0.9, P = 0.02) and blacks (odds ratio of $\widehat{or} = 0.60$: confidence interval 0.4 to 0.9, P = 0.02). Thus, women and black patients were less likely referred for cardiac catherization than men and white patients.

Associated Press:

"Doctors were 60 percent less likely to order cardiac catherization for women and blacks than for men and whites." Similarly reported by Washington Post, ABC Nightline (a nationally televised show), USA Today, CNN, New York Times, ABC, NBC, and Los Angeles Times.

Key calculations from the published analysis:

$$probability = P(white\ female\ referred\ to\ cardiac\ catherization) = 0.847$$

and

$$probability = P(white\ male\ referred\ to\ cardiac\ catherization) = 0.906$$

yield

$$relative\ risk\ ratio = \widehat{rr} = \frac{0.847}{0.906} = 0.935 \quad and$$

$$odds\ ratio = \widehat{or} = \frac{0.847 / 0.153}{0.906 / 0.094} = 0.574.$$

The conditions for an odds ratio to approximately measure risk are not fulfilled. The symptoms described in the interview are not sufficiently rare. Thus, creating an odds ratio that is an exceptionally misleading estimate of risk.

The journal editors reply to media:

"We take responsibility for the media's over interpretation of the article. We should not have allowed the use of the odds ratio in the abstract. As pointed out, risk ratios would have been clearer to readers…"

BOTTOM LINE: Why not compare 85% to 91%????

A brief review:

$$p = probability = P(disease\ among\ exposed\ individuals) and$$

$$P = probability = P(disease\ among\ unexposed\ individuals)$$

then

$$odds\ ratio = \widehat{or} = \frac{\dfrac{p}{1-p}}{\dfrac{P}{1-P}} \quad and \quad relative\ risk\ ratio = \widehat{rr} = \frac{p}{P}.$$

For the example, if $p = 0.847$ and $P = 0.906$, then

$$odds\ ratio = \widehat{or} = \frac{\dfrac{0.847}{1-0.847}}{\dfrac{0.906}{1-0.906}} \quad and \quad relative\ risk\ ratio = \widehat{rr} = \frac{0.847}{0.906}, yielding$$

$$odds\ ratio = \widehat{or} = 0.574\ and\ relative\ risk\ ratio = \widehat{rr} = 0.935.$$

The difference between an odds ratio and a relative risk ratio can be striking. For similar values $p = 0.9$ and $P = 0.8$, then

$$odds\ ratio = \frac{0.9/0.1}{0.8/0.2} = 2.25 \quad and \quad relative\ risk\ ratio = \frac{0.9}{0.8} = 1.12$$

or, the equivalent reciprocal values are:

$$odds\ ratio = \frac{0.8/0.2}{0.9/0.1} = 0.44 \quad and \quad relative\ risk\ ratio = \frac{0.8}{0.9} = 0.89.$$

Table 13.3 A ridiculously extreme example

	D	$\bar{D}$	total
exposed	20	1	21
unexposed	1	20	21
total	21	21	42

then, $\widehat{or} = \dfrac{20/1}{1/20} = 400$ and $\widehat{rr} = \dfrac{20/21}{1/21} = 20.$

A word to the wise:

The odds ratio and relative risk ratio have similar names, similar construction, play similar roles in assessing binary outcomes data and are often used interchangeably. However, as noted, they only produce similar results under specific conditions.

Reference: unpublished report (1994) to the US Veterans Administration (1988).

Reference: Schulman KA, Berlin JA, Harless W, et al. The effect of race and sex on physicians' recommendations for cardiac catheterization. *Journal of the American Medical Association*, 1999; 340(8): 618–26.

14

Odds ratio—two properties rarely mentioned

An often ignored property of a statistical description of data is that the observed results depend on the selection of the data summary. Data from a smoking and asbestos study of lung cancer mortality are an example.

INTERACTION: Smoking and asbestos exposure:

Excess mortality—smokers versus non-smokers:

Table 14.1 Age-adjusted lung cancer mortality rates per 100 000 persons among Canadian insulation workers (1987)

	smoker	non-smoker
exposed	601.6	58.4
not exposed	122.6	11.3

$$\text{exposed: rate difference} = 601.6 - 58.4 = 543.2$$

$$\text{not exposed: rate difference} = 122.6 - 11.3 = 111.3$$

and

$$\text{estimated excess mortality: difference} = 543.2 - 111.3 = 431.9.$$

Estimated odds ratio—smoker versus non-smoker:

$$\hat{or} = \frac{601.6 / 58.4}{122.6 / 11.3} = \frac{10.30}{10.85} = 0.95.$$

Interaction is a term given to failure of one variable (for example, smoking and non-smoking) to have the same relationship at the levels of a second variable (for example, exposed and not exposed). Excess lung

The Joy of Statistics: A Treasury of Elementary Statistical Tools and their Applications. Steve Selvin. © Steve Selvin 2019. Published in 2019 by Oxford University Press. DOI: 10.1093/oso/9780198833444.001.0001

cancer mortality between smokers and non-smokers is considerably different depending on exposure status, a strong interaction (excess mortality = 431.9). Comparison of the same mortality rates produces an odds ratio measure of influence of smoking exposure that yields only a small difference in lung cancer mortality, little indication of an interaction $(\hat{or} = 0.95)$. It is important to note that the extent of interaction between smoking and exposure status depends on the statistical measure used.

Two versions of the same lung cancer mortality data

interaction?—additive scale

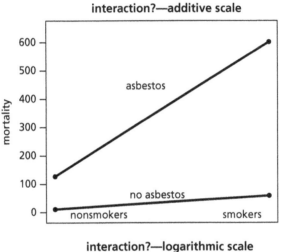

interaction?—logarithmic scale

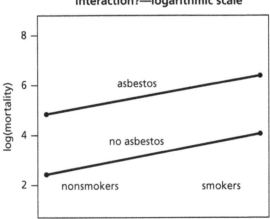

Two examples illustrating the simple but critical fact that interpretation of many statistics, such as the presence or absence of an interaction, frequently depends on choice of scale.

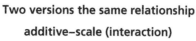

Two versions the same relationship
additive–scale (interaction)

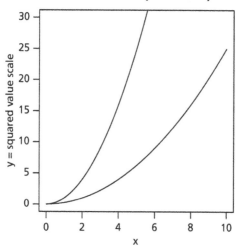

log–scale (no interaction)

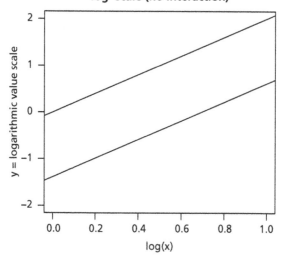

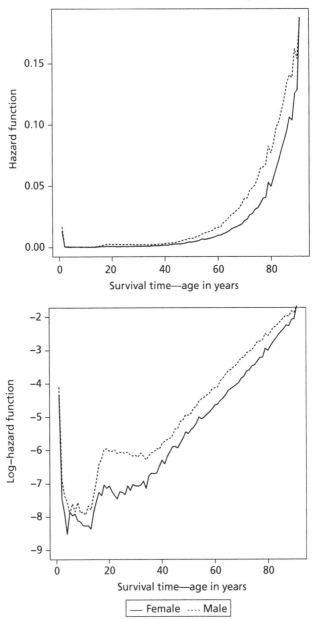

Variance of the odds ratio

An estimated odds ratio, like all estimates, becomes more precise as sample size increases. That is, the variance of the estimated value decreases. In the case of an odds ratio, a specific property of its variance makes it important to understand this relationship in detail.

Table 14.2 Breast cancer diagnosed (*D*) among military women who served in Vietnam (*exposed*) and who did not serve in Vietnam (*unexposed*) during the war years 1965–1973

	D	**$\bar{D}$**	total
exposed	170	3222	3392
unexposed	126	2912	3038
total	296	6134	6430

Estimated variance of the logarithm of the estimated odds ratio $\widehat{or} = 1.219$ is:

$$variance(log([\widehat{or}])) = \frac{1}{170} + \frac{1}{3222} + \frac{1}{126} + \frac{1}{2912} = 0.0145.$$

Ignoring women without a diagnosis of breast cancer $(\bar{D})$, the estimated variance is:

$$variance(log([\widehat{or}])) = \frac{1}{170} + \frac{1}{126} = 0.0138.$$

Including the 6134 non-cancer military women is necessary to estimate influence of exposure but produces no appreciable increase in precision. Therefore, in the case of these breast cancer data, more realistically, the "sample size" should be considered as closer to 170 + 126 = 296 and not 170 + 126 + 3222 + 2912 = 6430.

In general, the precision of an estimated log-odds ratio or odds ratio is extremely sensitive to small numbers of observations because of their considerable influence on the variance of the estimated value.

Reference: US mortality data—National Vital Statistical Reports, volume 64, Number 11, by Elizabeth Division Vital Statistics, 2015.

Reference: unpublished report to the US Veterans Administration (1988).

15

Percent increase—ratios?

"Debbie and Rich Hedding of Pittsfield Vt. were devastated when they lost two babies to neural tube defects (NTDs). When Debbie read about folic acid in the March of Dimes brochure, she was astonished when she learned about the role of folic acid in preventing NTDs. 'I was in tears by the time I finished reading the material. I couldn't believe that I could have reduced the risk of recurrence by 72 percent. I immediately began taking folic acid and telling every woman I could about it.' "

From the March of Dimes brochure (1998), the percentage decease is $(1 - 0.28) \times 100 = 72\%$.

From a study of neural tube defects and folic acid reported in the British medical journal *The Lancet*, odds ratio $= \hat{or} = 0.28$.

Aside:

For an additive scale, when $a - b = 10 - 5 = 5$ or $b - a = 5 - 10 = -5$, b is five less than a or a is five more than b. Both measures are different versions of the identical difference. For a multiplicative scale, when $a/b = 10/5 = 2$ or $b/a = 5/10 = 1/2$, a is twice the size of b or b is half the size of a. Again, both measures are different versions of the identical difference.

From another point of view:

Table 15.1 Comparison of two example tables

	exposed	not exposed
male	10	3
female	2	3

Table 15.2 The same data

	exposed	not exposed
female	2	3
male	10	3

The Joy of Statistics: A Treasury of Elementary Statistical Tools and their Applications. Steve Selvin. © Steve Selvin 2019. Published in 2019 by Oxford University Press. DOI: 10.1093/oso/9780198833444.001.0001

$$\text{yields the odds ratio } \widehat{or} = \frac{2/3}{10/3} = 1/5 = 0.2 .$$

As required, the odds ratio $\widehat{or} = 0.2$ is equivalent to $\widehat{or} = 5.0$ which is another version of the identical ratio. That is, the same data necessarily produce the same association in both tables. In general, the odds ratio $\widehat{or}$ and $1/\widehat{or}$ are different versions of the identical association. A percent decrease such as 72 percent is not relevant to interpreting values compared on a ratio scale.

Debbie would have been even more upset if the calculation was done correctly. That is, $\widehat{or} = 1/\widehat{or} = 1/0.28 = 3.6$, sometimes referred to as more than a 3.6-fold increase or decrease.

One last note: the logarithm of an odds ratio produces a symmetric measure of association. Specifically, $-log(\widehat{or}) = log(1/\widehat{or})$ making $-log(0.28) = log(1/0.28) = 1.27$. A percent increase of 72 percent is

$$\frac{1.27 - 1.0}{1.27} \times 100 = 21.3$$

which is mathematically correct but not useful statistically as measure of association.

Examples from six medical journals:

Journal National Cancer Institute
odds ratio = 0.80 stated as 20 percent reduction

Obstetrics and Gynecology
odds ratio = 1.20 stated as 20 percent increase

New England Journal of Medicine
odds ratio = 0.40 stated as 60 percent reduction

Cancer Epidemiology
odds ratio = 1.35 stated as 35 percent increase

Annals of Epidemiology
odds ratio = 0.61 stated as 39 percent reduction

Cancer Causes and Control
odds ratio = 0.70 stated as 30 percent reduction

A letter to the editor

Defending criticism of an analysis based on a relative risk measure that appeared in the *Journal of Cancer, Epidemiology Biomarkers and Prevention*, the authors replied:

"We disagree with the critics that correct interpretation of the RR [relative risk] measure is the inverse of the estimate when the RR is <1.0. We know of no statistical method suggesting that the RR or odds ratios should be interpreted as the inverse of the calculated measure of association. Your letter [recently published by the journal] contains no reference to support this proposition."

16

Diagnostic tests—assessing accuracy

Headline—"Parents Add Drug Tests to Shopping List" from a CVS and Wall Street Journal newspaper article (1997).

The newspaper report describes an increasing trend in purchasing home drug testing kits. The market is clearly parents who worry, "Is my teenager doing drugs?" Kits costing between 60 and 140 dollars are sold by drugstore chains and stores such as Walmart.

Television evening network news explored the same issue with a typical interview (a sample of size one):

An evening news (NBC) interview of members of a Kentucky family:

Interviewer: What do you think of your son's positive test indicated by a home testing kit to detect smoking marijuana?

Father: We are very happy with the results. We have had a useful and productive discussion with our son.

Mother: We certainly do not want our children to become involved in illegal drug use.

Teenage son [creditably]: I can understand my parents' concern. However, I have never smoked marijuana and I plan to never try it.

An important question certainly arises: what is the likelihood the home drug test reflects a false positive result? That is, the test indicates the son has smoked marijuana when in fact he has not.

A general, simple, and often used approach to this question is described by a 2×2 table created to assess a positive/negative performance of a test to identify presence/absence of an outcome. For example, a test for presence/absence of drugs or disease. For the following discussion, presence/absence of disease is described to simplify terminology. However, assessment of accuracy of a diagnostic test applies to many different binary presence/absence outcomes. Three examples are: presence/absence of a virus in mosquitoes, presence/absence of use of illegal narcotics, and presence/absence of performance-enhancing drugs in athletes.

The Joy of Statistics: A Treasury of Elementary Statistical Tools and their Applications. Steve Selvin. © Steve Selvin 2019. Published in 2019 by Oxford University Press. DOI: 10.1093/oso/9780198833444.001.0001

Three probabilities that are key to issues central to a diagnostic test are displayed in a 2 × 2 table of test results. The probabilities are given specific names:

1. **Sensitivity** is defined as the probability a test identifies an individual as positive for a disease among those who have the disease, denoted $p[sensitivity]$ and estimated by $a/(a + c)$.
2. **Specificity** is defined as the probability a test identifies an individual as negative for a disease among those who do not have the disease, denoted $p[specificity]$ and estimated by $d/(b + d)$.
3. **False positive** is defined as the probability of a positive test for an individual who does not have the disease and estimated by $b/(a + b)$.

The probabilities sensitivity and specificity displayed in a 2 × 2 table of n test results where P denotes prevalence of disease:

Table 16.1 Four possible outcomes of a binary test

Test	Disease present	Disease absent	Total
positive	$a = nP \times p[sensitivity]$	$b = n(1 - P) \times (1 - p[specificity])$	$a + b$
negative	$c = nP \times (1 - p[sensitivity])$	$d = n(1 - P) \times p[specificity]$	$c + d$
total	$a + c = nP$	$b + d = n(1 - P)$	n

Three hypothetical probabilities illustrate calculation of the probability of a false positive test, where

P = prevalence of disease = 0.10, $p[sensitivity]$ = 0.95 and $p[specificity]$ = 0.90.

The calculation is straightforward. Table for any sample size (n), for example, $n = 2000$:

Table 16.2 Illustration of four possible outcomes of a binary test

Test	Disease present	Disease absent	total
positive	$a = 2000(0.10)[0.95] = 190$	$b = 2000(0.90)[0.10] = 180$	$a + b = 370$
negative	$c = 2000(0.10)[0.05] = 10$	$d = 2000(0.90)[0.90] = 1620$	$c + d = 1630$
total	$a + c = nP = 200$	$b + d = n(1 - P) = 1800$	$n = 2000$

Therefore, the estimated probability of a false positive test is P(*false positive*) = $b/(a + b)$ = 180/370 = 0.486. Thus, in the case of the Kentucky teenager, for prevalence of marijuana use P = 0.10, p[*sensitivity*] = 0.95 p[*specificity*] = 0.90 and the probability of a misleading false positive test result is close to 0.5. The calculation suggests a rather expensive test that is not very effective.

Another example:

When athletes are tested for presence of illegal drugs after a competition, high sensitivity means drug users will likely be correctly identified. Similarly, high specificity means non-drug users will likely be correctly identified. High sensitivity and specificity are obvious properties of an accurate test. Less obvious is the fact that performance of a binary test, measured in terms of the likelihood of a false positive outcome, is highly influenced by prevalence of the outcome (P). As prevalence decreases, the probability of a false positive test increases, dramatically increases.

To repeat, decreasing prevalence causes increasing likelihood of a false positive test. Counts of values from four 2 × 2 tables (rows) used to calculate the probability of a false positive test result for four prevalence frequencies (again, denoted P) with p[*sensitivity*]= 0.95 and p[*specificity*] = 0.90 are:

Table 16.3 Four example probabilities of a false positive outcome

P	a	b	c	d	false positive*
0.50	950	100	50	900	0.095
0.10	190	180	10	1620	0.486
0.05	95	190	5	1710	0.667
0.01	19	198	1	1782	0.912

* = as noted, false positive probability = $b/(a + b)$

When an outcome is rare, such as prevalence of HIV, a positive test is treated as evidence that more comprehensive, extensive, and frequently more expensive tests are necessary because of the likelihood of a false positive result. Also noted, as prevalence (P) decreases toward zero, probability of a false positive test rapidly increases toward 1.0. When $P = 0$, all positive tests are false.

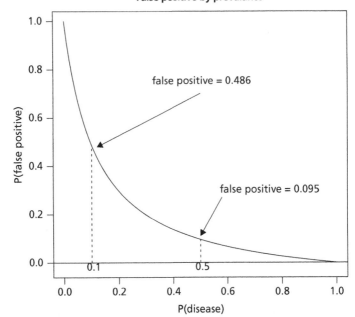

17

Regression to the mean— father/son data

A dictionary definition of the word regression:

a return to a former or less developed state.

An important phenomenon arises from data consisting of pairs of related values called "regression to the mean." The defining property emerged (circa 1900) when Sir Francis Galton observed tall fathers tended to have shorter sons and short fathers tended to have taller sons. Statisticians Karl Pearson and Alice Lee collected heights of 1078 pairs of British fathers and sons.

The table and plot from their original data unequivocally support Galton's observation.

Table 17.1 Comparison of father/son heights*

son	father	difference
71.6	74.9	−3.31
71.5	72.9	−1.41
70.6	71.5	−0.89
70.0	70.1	−0.12
67.0	64.3	2.69
66.2	62.7	3.56
65.4	60.8	4.62
64.7	59.4	5.30

* = mean values of heights in inches based on n = 1078 father/son pairs.

The plotted diagonal line indicates exactly equal father and son heights. Clearly the observed distribution of father/son pairs differs from this perhaps expected relationship (slope = 1.0). An estimated

The Joy of Statistics: A Treasury of Elementary Statistical Tools and their Applications. Steve Selvin. © Steve Selvin 2019. Published in 2019 by Oxford University Press. DOI: 10.1093/oso/9780198833444.001.0001

summary line describing a linear father/son relationship clearly indicates fathers taller than average likely have shorter sons and fathers shorter than average likely have taller sons. More technically, the line describing father/son heights has an estimated slope of close to 0.5 which occurs because a son's height is frequently less than his tall father's height and a son's height is frequently taller than his short father's height, called *regression to the mean*.

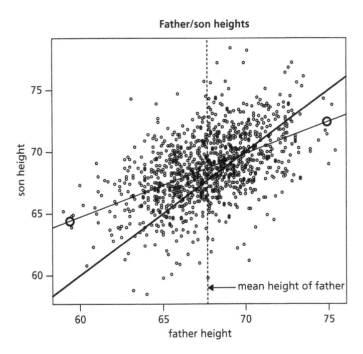

Father/son heights

An example:

From the plot (two bold circles) and table, tall fathers (greater than 74 inches—table first row) have a mean height of 74.9 inches and their sons have mean height of 71.6 inches (3.3 inches shorter). Similarly, short fathers (less than 60 inches—table last row) have a mean height of 59.4 inches and their sons have mean height of 64.7 inches (5.3 inches taller).

Recognition of this father/son relationship caused concern early in 20th century England. It was thought that if this trend continued over a number of generations, eventually British men could become essentially the same height. Other genetic traits such as intelligence, strength, weight, and various inherited skills could also eventually regress to a single level. Thus, the name "regression to the mean." Perhaps a concern of the British upper class was that social extremes would similarly regress to a homogeneous society. However, as knowledge of genetics and statistics became more sophisticated, it became apparent variation among heights remains the same every generation. That is, the genetics that determine height are not affected by regression to the mean observed in the previous generation.

The "Sports Illustrated jinx" is an example of regression to the mean. First, what is the Sports Illustrated jinx? It has been pointed out that after "the athlete of the year" picture appears on the cover of Sports Illustrated magazine, the following year he or she rarely lives up to the previous performance. That is, outstanding performances are typically followed by less outstanding performances caused simply by the fact that it is difficult to repeat an exceptional performance. The word "exceptional" is key. Chance is an important element in an outstanding performance. The occurrence of "lucky" conditions that typically played a role in an outstanding performance are not likely to occur again. Injuries, weather, different opponents, and a long list of other possibilities reduce the chance of a repeat athletic performance. It is this random element between consecutive occurrences that causes regression to the mean. Simply, rare events are not likely to occur twice. Thus, "when you are down, the likely way is up" or "when you are up, the likely way is down."

The term *regression analysis* refers to frequently used statistical techniques. When a sampled observation is made up of components, called a multivariate observation, it is often important to explore the relative importance of these components. For example, an infant's birth weight is a multivariate observation resulting from a large number of influences. A few examples are: maternal age, weight, and ethnicity, as well as many other factors that may also be important. A regression analysis provides a statistical exploration of the relative importance of these components in determining birth weight. The statistical term *regression* evolved from the Galton/Pearson/Lee recognition of regression to the mean but when applied to modern statistical analysis it is not used in the sense of returning to a former less developed state. It has become simply the name given a specific kind of analysis.

Reference: complete original data available University of California, Berkeley mathematical sciences.

18

Life table—a summary of mortality experience

Two 17th century English scholars, Edmund Halley and John Grunt, independently developed a remarkable description of human mortality patterns called a *life table*. Construction of a life table requires a simple approximate relationship. A life table is derived from current mortality data treated as if the data were collected longitudinally over a lifetime. More specifically, newborn infants are treated as if the present pattern of mortality for the next 100 years does not change. From yet another prospective, a newborn infant is assumed to have exactly the same mortality risk in 60 years experienced by a person 60 years old during the year the infant was born. Clearly, mortality patterns change over time. Nevertheless, changes in human mortality patterns are generally small, making life table summaries useful for short-term predictions and excellent for comparing current mortality patterns among groups and populations.

Table 18.1 Mechanics of life table construction (US vital records, 2010)

1. Age: 0, 1, 2, 3,···, 100 years (denoted x):

   age interval x to $x+1$ (length = one year)

2. Number alive at age x years (denoted l_x):

   $l_{60} = 88,770$ with $l_0 = 100,000$ (an artificial population)

   $l_0, l_1, l_2, \cdots, l_{60} = 88,770, \cdots, l_{100^+}$

3. Deaths between age x to $x+1$ years (denoted dx):

   Ages 60 to 61 and $d_{60} = 788$

   $d_0, d_1, d_2, \cdots, d_{60} = 788, \cdots, d_{100^+}$

The Joy of Statistics: A Treasury of Elementary Statistical Tools and their Applications. Steve Selvin. © Steve Selvin 2019. Published in 2019 by Oxford University Press. DOI: 10.1093/oso/9780198833444.001.0001

4. Probability of death between age x to $x + 1$ years (denoted q_x):

$$q_x = \frac{d_x}{l_x} \quad \text{and} \quad q_{60} = \frac{778}{88,770} = 0.00876$$

$q_0, q_1, q_2, \cdots, q_{60} = 0.00876, \cdots, q_{100^+}$

5. Total person-years lived from age x to $x + 1$ years (denoted L_x):

$x = 60, l_{60} = 88,770 \quad \text{and} \quad l_{61} = 87,992$

$d_{60} = l_{60} - l_{61} = 88,770 - 87,992 = 778$

then, approximate total person-years lived from age x is:

$$L_x = l_x - \frac{1}{2}d_x \text{ and } L_{60} = l_{60} - \frac{1}{2}d_{60} = 88,770 - \frac{1}{2}(778)$$
$$= 88,381 \text{ person-years}$$

The expression $\frac{1}{2}d_x = 389$ is approximately the person-years lived by those who died during interval ages 60 to 61.

6. Total person-years lived beyond age x (denoted T_x):

$T_x = L_x + L_{x+1} + L_{x+2} + \cdots + L_{100^+}$

$T_{60} = 88,381 + 87,578 + 86,725 + \cdots + 2823 = 2,046,759$ person-years

$T_0 = 99,465 + 99,366 + \cdots + 2823 = 7,866,027$ person-years

7. Expected (mean) time alive beyond age x (denoted e_x):

$$e_{60} = \frac{T_{60}}{l_{60}} = \frac{2,046,759}{88,770} = 23.1 \text{ years}$$

$$e_0 = \frac{T_0}{l_0} = \frac{7,866,027}{100,000} = 78.7 \text{ years}$$

8. Life table mortality rate for a person 60 years old during the next year of life is:

$$rate_{60/61} = \frac{d_{60}}{\frac{1}{2}\delta(l_{60} + l_{61})} = \frac{778}{\frac{1}{2}(1)(88,770 + 87,992)} = 0.009$$

$$= 8.8 \text{ deaths per 1000 person-years}$$

Table 18.2 Total population: United States, 2010

Age	q_x	l_x	d_x	L_x	T_x	e_x
0–1	0.006123	100,000	612	99,465	7,866,027	78.7
1–2	0.000428	99,388	43	99,366	7,766,561	78.1
2–3	0.000275	99,345	27	99,331	7,667,195	77.2
3–4	0.000211	99,318	21	99,307	7,567,864	76.2
4–5	0.000158	99,297	16	99,289	7,468,556	75.2
5–6	0.000145	99,281	14	99,274	7,369,267	74.2
6–7	0.000128	99,267	13	99,260	7,269,993	73.2
7–8	0.000114	99,254	11	99,249	7,170,733	72.2
8–9	0.000100	99,243	10	99,238	7,071,484	71.3
9–10	0.000087	99,233	9	99,229	6,972,246	70.3
10–11	0.000079	99,224	8	99,220	6,873,017	69.3
11–12	0.000086	99,216	9	99,212	6,773,797	68.3
12–13	0.000116	99,208	12	99,202	6,674,585	67.3
13–14	0.000175	99,196	17	99,188	6,575,383	66.3
14–15	0.000252	99,179	25	99,167	6,476,195	65.3
15–16	0.000333	99,154	33	99,138	6,377,028	64.3
16–17	0.000412	99,121	41	99,101	6,277,891	63.3
17–18	0.000492	99,080	49	99,056	6,178,790	62.4
18–19	0.000573	99,032	57	99,003	6,079,734	61.4
19–20	0.000655	98,975	65	98,942	5,980,731	60.4
20–21	0.000744	98,910	74	98,873	5,881,789	59.5
21–22	0.000829	98,836	82	98,795	5,782,916	58.5
22–23	0.000892	98,754	88	98,710	5,684,120	57.6
23–24	0.000925	98,666	91	98,621	5,585,410	56.6
24–25	0.000934	98,575	92	98,529	5,486,789	55.7
25–26	0.000936	98,483	92	98,437	5,388,260	54.7
26–27	0.000943	98,391	93	98,344	5,289,824	53.8
27–28	0.000953	98,298	94	98,251	5,191,479	52.8
28–29	0.000971	98,204	95	98,157	5,093,228	51.9
29–30	0.000998	98,109	98	98,060	4,995,071	50.9
30–31	0.001029	98,011	101	97,961	4,897,011	50.0
31–32	0.001063	97,910	104	97,858	4,799,051	49.0
32–33	0.001099	97,806	108	97,752	4,701,193	48.1
33–34	0.001137	97,699	111	97,643	4,603,440	47.1
34–35	0.001180	97,587	115	97,530	4,505,797	46.2
35–36	0.001235	97,472	120	97,412	4,408,267	45.2
36–37	0.001302	97,352	127	97,289	4,310,855	44.3
37–38	0.001377	97,225	134	97,158	4,213,567	43.3
38–39	0.001461	97,091	142	97,020	4,116,408	42.4
39–40	0.001557	96,949	151	96,874	4,019,388	41.5
40–41	0.001663	96,798	161	96,718	3,922,514	40.5
41–42	0.001793	96,637	173	96,551	3,825,796	39.6

(*Continued*)

Table 18.2 Continued

Age	q_x	l_x	d_x	L_x	T_x	e_x
42–43	0.001962	96,464	189	96,370	3,729,245	38.7
43–44	0.002177	96,275	210	96,170	3,632,875	37.7
44–45	0.002423	96,065	233	95,949	3,536,705	36.8
45–46	0.002676	95,833	256	95,704	3,440,756	35.9
46–47	0.002931	95,576	280	95,436	3,345,052	35.0
47–48	0.003205	95,296	305	95,143	3,249,616	34.1
48–49	0.003505	94,990	333	94,824	3,154,473	33.2
49–50	0.003830	94,658	363	94,476	3,059,649	32.3
50–51	0.004177	94,295	394	94,098	2,965,173	31.4
51–52	0.004535	93,901	426	93,688	2,871,075	30.6
52–53	0.004903	93,475	458	93,246	2,777,386	29.7
53–54	0.005284	93,017	491	92,771	2,684,140	28.9
54–55	0.005684	92,526	526	92,263	2,591,369	28.0
55–56	0.006117	92,000	563	91,718	2,499,106	27.2
56–57	0.006589	91,437	603	91,136	2,407,388	26.3
57–58	0.007095	90,834	644	90,512	2,316,253	25.5
58–59	0.007626	90,190	688	89,846	2,225,741	24.7
59–60	0.008180	89,502	732	89,136	2,135,895	23.9
60–61	0.008767	88,770	778	88,381	2,046,759	23.1
61–62	0.009397	87,992	827	87,578	1,958,378	22.3
62–63	0.010085	87,165	879	86,725	1,870,800	21.5
63–64	0.010863	86,286	937	85,817	1,784,075	20.7
64–65	0.011758	85,348	1004	84,847	1,698,258	19.9
65–66	0.012810	84,345	1080	83,805	1,613,411	19.1
66–67	0.014011	83,264	1167	82,681	1,529,606	18.4
67–68	0.015290	82,098	1255	81,470	1,446,925	17.6
68–69	0.016601	80,843	1342	80,172	1,365,455	16.9
69–70	0.018005	79,501	1431	78,785	1,285,283	16.2
70–71	0.019548	78,069	1526	77,306	1,206,499	15.5
71–72	0.021294	76,543	1630	75,728	1,129,192	14.8
72–73	0.023275	74,913	1744	74,041	1,053,464	14.1
73–74	0.025528	73,169	1868	72,236	979,423	13.4
74–75	0.028061	71,302	2001	70,301	907,188	12.7
75–76	0.030820	69,301	2136	68,233	836,886	12.1
76–77	0.033775	67,165	2268	66,031	768,654	11.4
77–78	0.037252	64,896	2418	63,688	702,623	10.8
78–79	0.041136	62,479	2570	61,194	638,935	10.2
79–80	0.045411	59,909	2721	58,549	577,741	9.6
80–81	0.050146	57,188	2868	55,754	519,193	9.1
81–82	0.055445	54,321	3012	52,815	463,438	8.5
82–83	0.061272	51,309	3144	49,737	410,624	8.0
83–84	0.067764	48,165	3264	46,533	360,887	7.5

(*Continued*)

Table 18.2 Continued

Age	q_x	l_x	d_x	L_x	T_x	e_x
84–85	0.075818	44,901	3404	43,199	314,354	7.0
85–86	0.085319	41,497	3540	39,727	271,155	6.5
86–87	0.094975	37,956	3605	36,154	231,429	6.1
87–88	0.105525	34,351	3625	32,539	195,275	5.7
88–89	0.117007	30,726	3595	28,929	162,736	5.3
89–90	0.129450	27,131	3512	25,375	133,807	4.9
90–91	0.142873	23,619	3375	21,932	108,432	4.6
91–92	0.157280	20,245	3184	18,653	86,500	4.3
92–93	0.172661	17,061	2946	15,588	67,847	4.0
93–94	0.188988	14,115	2668	12,781	52,259	3.7
94–95	0.206214	11,447	2361	10,267	39,478	3.4
95–96	0.224274	9087	2038	8068	29,211	3.2
96–97	0.243080	7049	1713	6192	21,144	3.0
97–98	0.262527	5335	1401	4635	14,951	2.8
98–99	0.282492	3935	1112	3379	10,316	2.6
100+	0.302838	2823	855	2396	6937	2.5

Table 18.3 US expected years of life (e_0) for the 20th century

years	1900	1920	1940	1960	1980	2000	increase
males	46.6	54.5	62.1	67.4	70.0	71.3	24.7
females	48.7	55.6	66.6	74.1	78.1	79.8	31.1
difference	2.1	1.1	4.5	6.7	8.1	8.5	6.4

Many reasons for increasing expected lifetime (approximately 28 years) in the United States over the last century are obvious and well known. However, reasons for increasing difference in expected lifetimes between females and males are not as well understood.

Table 18.4 Values of expected lifetime e_0 (both sexes) for a few selected countries (2010)

Japan = 83.7, Sweden = 82.4, United States = 78.7, Mexico = 76.7, China = 78.6, Brazil = 74.9, Egypt = 70.9, Congo = 59.8, Somali = 55.0, Sierra Leone = 50.1

rank 1: Swiss (women) = 86.8 and rank 183: Sierra Leone (men) = 49.3

The approximate relationship between a life table mortality rate and probability of death for year x is

$$rate_x = \frac{d_x}{\frac{1}{2}\delta(l_x + l_{x+\delta})} = \frac{d_x}{\delta(l_x - \frac{1}{2}d_x)} \approx \frac{d_x}{\delta l_x} \approx \frac{q_x}{\delta}$$

because approximately

$$l_x - \frac{1}{2}d_x \approx l_x \quad \text{since typically} \quad l_x \gg d_x.$$

For example, for age $x = 60$, then $l_{60} = 88,770$ and the considerably smaller value is $\frac{1}{2}d_{60}$ is 389.

Thus, mortality rates approximately equal the probability of death divided by the length of the time interval under consideration (denoted δ). In symbols, $rate_x \approx q_x/\delta$ for the interval x to $x + \delta$.

For age interval 60 to 70 years:

$\delta = 10$ years, $l_{60} = 88,770$, and $l_{70} = 78,069$, then the life table mortality rate is:

$$rate_{60/70} = \frac{d_{70}}{\frac{1}{2}\delta(l_{60} + l_{70})} = \frac{10,701}{\frac{1}{2}(10)(88,770 + 78,069)}$$

$$\approx 12.8 \text{ per 1000 person-years}$$

The approximate probability of death follows as $q_{60/70} = rate_{60/70} \times \delta = 0.0128 \times 10 = 0.128$ for age interval 60 to 70 years.

The probability of surviving beyond a specific age x is calculated directly for life table values l_x. The probability of surviving beyond age x (denoted P_x) is called a survival probability and

$$survival\ probability = P_x = \frac{number\ alive\ at\ age\ x}{number\ at\ age\ 0} = \frac{l_x}{l_0} = \frac{l_x}{100,000}.$$

For example, for age 60, the survival probability is $P_{60} = 88,700/100,000 = 0.887$ that a newborn infant will survive beyond age 60. Remarkably, the survival curve is close to level (constant probability), remaining above 0.90 until about age 60. At age 60, survival probabilities dramatically decline yielding probability of surviving beyond age 100 years as $2823/100,000 = 0.028$.

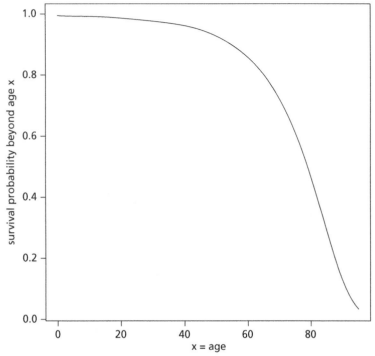

A plot describing the probability of death by age is, not surprisingly, an "inverse" of the survival probability plot. This description of a mortality pattern is usually called a *hazard function*. For one year intervals ($\delta = 1$), a hazard function is closely approximated by the life table age-specific probabilities of death (q_x). As predicted from the survival curve, the hazard function is close to constant from ages 0 to 60 with a sharp increase starting at about age 60 years.

Television "health" shows, commercials, advertisements, and a variety of other sources frequently refer to lifetime probability of contracting a disease. For example, something like "during a women's lifetime she will be diagnosed with breast cancer with a probability of 1 in 12." This probability has little use. The probability of disease over a lifetime obviously varies from close to 0 among the very young and close to 1.0 among the very old. No single probability of

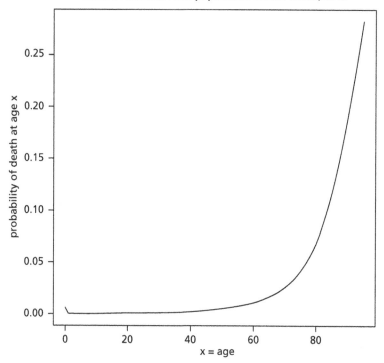

Hazard Function: Total population: United States, 2010

Table 18.5 A three century comparison*: Probability of surviving beyond age x years

century	x = 10	x = 20	x = 30	x = 40	x = 50	x = 60	x = 70	x = 80
17th	0.54	0.34	0.21	0.14	0.08	0.05	0.02	0.01
20th	0.99	0.99	0.98	0.97	0.95	0.94	0.89	0.78

* = two locations with populations: London = 527,000 (approximate) and Berkeley, California = 116,768 (US census)

Life table survival probabilities

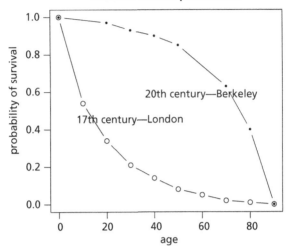

death or disease usefully reflects risk without a specific consideration of age. In other words, age is always fundamental to any description of risk.

As might be expected, years of remaining lifetime (*ex*) decrease almost linearly. That is, every year lived reduces expected additional years lived about one year. The mean decrease from the US life table is 1.02 per year lived.

Note: close to half the infants born in 17th century London died before the age of ten.

Reference: National Vital Statistical Reports, volume 64, Number 11, by Elizabeth Division Vital Statistics, 2010.

19

Coincidence—a statistical description

Dictionary definition of coincidence:

A remarkable concurrence of events or circumstances without apparent causal connection.

When two rare events simultaneously occur, a usually unsuccessful search for an explanation begins. In a statistical context, a coincidence is clearly explored in terms of rigorous calculations.

Two distinct types of coincidences are:

Type 1:

Simultaneous occurrences of two improbable unrelated events. For example, winning two separate lotteries on the same day.

Type 2:

Occurrence of a second improbable event matching an already existing improbable event. For example, winning lottery numbers selected that are identical to the winning numbers selected the previous week.

The birthday problem

A traditional problem concerning occurrence of birthdays illustrates the difference between type 1 and type 2 coincidences. Consider a group of individuals with equally likely birthdays. In fact, birthdays throughout a year are only approximately equally likely.

Computation of two relevant probabilities:

Coincidence type 1:

Among n persons, what is the probability two randomly chosen individuals have the same birthday day?

$$P(\text{one pair among two persons}) = 1 - \left(1 - \frac{1}{365}\right) = 0.003$$

$$P(\text{one pair among three persons}) = 1 - \left(1 - \frac{1}{365}\right) \times \left(1 - \frac{2}{365}\right) = 0.008$$

$$P(\text{one pair among four persons}) = 1 - \left(1 - \frac{1}{365}\right) \times \left(1 - \frac{2}{365}\right) \times \left(1 - \frac{3}{365}\right) = 0.0016$$

The Joy of Statistics: A Treasury of Elementary Statistical Tools and their Applications. Steve Selvin. © Steve Selvin 2019. Published in 2019 by Oxford University Press. DOI: 10.1093/oso/9780198833444.001.0001

$$P(\text{one pair among } n \text{ persons}) = 1 - \left(1 - \frac{1}{365}\right) \times \left(1 - \frac{2}{365}\right) \times \left(1 - \frac{3}{365}\right) \times \cdots \times \left(1 - \frac{n-1}{365}\right).$$

Coincidence type 2:

Among n persons, what is the probability a randomly selected individual matches a specific birthday?

$$P(\text{matching a specific birthday chosen among } n \text{ individuals}) = 1 - \left(\frac{365-1}{365}\right)^{n}.$$

Table 19.1 Example coincidence probabilities

persons (n)	1	3	5	7	9	11	13	15
type 1	—	0.008	0.027	0.056	0.095	0.141	0.194	0.253
type 2	0.003	0.008	0.014	0.019	0.024	0.030	0.035	0.040

persons (n)	17	19	21	23	25	27	29	31
type 1	0.315	0.379	0.444	**0.507**	0.569	0.627	0.681	0.730
type 2	0.046	0.051	0.056	**0.061**	0.066	0.071	0.0076	0.082

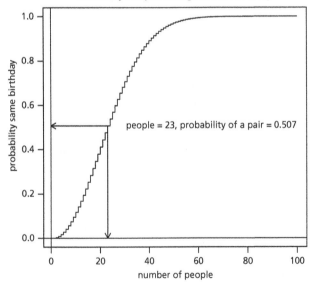

Probability of a pair having the same birthday

people = 23, probability of a pair = 0.507

Probability a person has a specific birthday

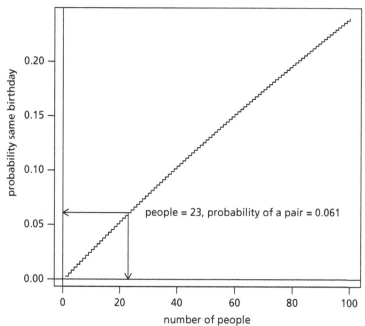

The two types of coincidences generate considerably different probabilities of matching birthdays. A typical value often chosen for comparison is 23 individuals ($n = 23$). In the first case, the probability a random pair have the same birthday (type 1) is 0.507. In the second case, the probability a randomly selected person matches a specified birthday (type 2) is the somewhat surprising value of 0.061.

For a group of 50 individuals, the probability of a randomly matching pair (type 1) is essentially a sure thing, or, more formally, $P(matching\ pair)$ ≈ 1.0. Matching a specific birthday (type 2) is 0.128.

Among six winners of the Massachusetts lottery, one player had two winning tickets and received 1/3 of the prize money. Sometimes a coincidence is not a coincidence. The player bought his ticket containing his lucky numbers from a lottery machine. As he walked away, he noticed one of his lucky numbers was wrong. He returned to the machine and again entered his lucky numbers. However, he made the same mistake again, producing two "wrong" and identical winning lottery tickets.

20

Draft lottery numbers (1970)

The 1970 United States Army draft began with randomly assigning numbers from 1 to 366 to each day of the year (1970 was a leap-year). Then, to fill the military draft quota from available men (essentially 18 year-olds), a specific set of numbers was selected (for example, 1 to 33). Men born on days with these selected numbers became eligible to be drafted. The purpose of the lottery was to notify those men who would be eligible to be drafted (low draft numbers, for the example, less than 34) and those who would not be eligible (remaining higher draft numbers). The set of randomly selected numbers assigned to 366 days for the 1970 draft:

305	159	251	215	101	224	306	199	194	325	329	221	318	238	17
121	235	140	58	280	186	337	118	59	52	92	355	77	349	164
211	86	144	297	210	214	347	91	181	338	216	150	68	152	4
89	212	189	292	25	302	363	290	57	236	179	365	205	299	285
108	29	267	275	293	139	122	213	317	323	136	300	259	354	169
166	33	332	200	239	334	265	256	258	343	170	268	223	362	217
30	32	271	83	81	269	253	147	312	219	218	14	346	124	231
273	148	260	90	336	345	62	316	252	2	351	340	74	262	191
208	330	298	40	276	364	155	35	321	197	65	37	133	295	178
130	55	112	278	75	183	250	326	319	31	361	357	296	308	226
103	313	249	228	301	20	28	110	85	366	335	206	134	272	69
356	180	274	73	341	104	360	60	247	109	358	137	22	64	222
353	209	93	350	115	**279**	188	327	98	190	227	187	27	153	172
23	67	303	289	88	270	287	193	111	45	261	145	54	114	168
48	106	21	324	142	307	198	102	44	154	141	311	344	291	339
116	36	286	245	352	167	61	333	11	225	161	49	232	82	6
8	184	263	71	158	242	175	1	113	207	255	246	177	63	204
160	119	195	149	18	233	257	151	315	359	125	244	202	24	87
234	283	342	220	237	72	138	294	171	254	288	5	241	192	243
117	201	196	176	7	264	94	229	38	79	19	34	348	266	310
76	51	97	80	282	46	66	126	127	131	107	143	146	203	185

The Joy of Statistics: A Treasury of Elementary Statistical Tools and their Applications. Steve Selvin. © Steve Selvin 2019. Published in 2019 by Oxford University Press. DOI: 10.1093/oso/9780198833444.001.0001

156	9	182	230	132	309	47	281	99	174	129	328	157	165	56
10	12	105	43	41	39	314	163	26	320	96	304	128	240	135
50	13	277	284	248	15	42	331	322	120	70	53	162	95	84
173	78	123	16	3	100									

Examples: January 1 = 305, July 4 = 279 and December 30 = 3, therefore men born on December 30 became eligible to be drafted. Question: Were the selected lottery numbers random?

Answer:

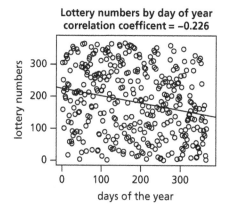

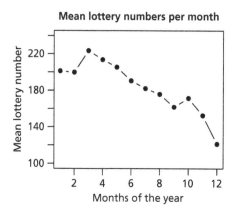

Graphical assessments unequivocally answer the question. The estimated summary line, months by corresponding monthly mean lottery numbers, tells the story. Low numbers occurred less frequently later in the year.

The selection of the 1970 lottery numbers was televised. A female military officer drew numbers from a barrel that had been rotated to mix the numbers. The graphical results are consistent with placing ordered numbers in the barrel and insufficiently rotating it to produce a random mix.

Reference: Draft Lottery (1969). Available at https://en.wikipedia.org/wiki/Draft_lottery_(1969) (accessed 3 December 2018).

21

Lotto—how to get in and how to win

From a book written by lotto expert Gail Howard:

> In a lotto game where you select six numbers out of the range 1 to 40, the ideal total of those numbers would be 120. In 6/40 Lotto, the mid-number is 20. The greatest percentage of possible six number combinations in the 6/40 game add up to 120. There is only one six number combination that adds to 21 (1+2+3+4+5+6 = 21) and only one six-number combination that adds to 225 (35+36+37+38+39+40 = 225). These two combinations are at the extreme tails of the bell curve [normal distribution], making it virtually impossible for these numbers to be drawn as a winning combination. Numbers that add to 120 are more frequent.

Summary:

> mid-number = 20,
> mean value = median = 20.5,
> total number of different six-number combinations = 3,383,380,
> probability of winning is one in 3,383,380 = 0.0000002605,
> total number of different values that add to 120 is 54,946 (1.46%),
> cost of the book = 2.99 dollars.

Everything Gail Howard wrote is more or less correct. However, nothing in her suggested strategy is relevant to winning. A lotto winner is determined by six randomly chosen numbers not their sum. Steve Jobs, an original founder of Apple Corporation, said when he bought a lottery ticket he chose numbers 1, 2, 3, 4, 5, and 6. He explained, any other six numbers have the same chance of winning as any six numbers [regardless of their sum] but the numbers 1–6 have the advantage of being easily remembered.

Newspaper headline: "A Stunning Coincidence."

The Bulgarian national 6/42-lottery (six random numbers chosen from 1 to 42) in September of 2009 produced the same six numbers (4, 15, 23, 24, 35, and 42) on consecutive weekly lotteries. The coincidence

The Joy of Statistics: A Treasury of Elementary Statistical Tools and their Applications. Steve Selvin. © Steve Selvin 2019. Published in 2019 by Oxford University Press. DOI: 10.1093/oso/9780198833444.001.0001

drew huge internet attention, world news coverage, and considerable speculation. The Bulgarian authorities promised to investigate.

Issues underlying the "stunning" coincidence become clear and calculations of probabilities straightforward when a distinction is made between two rare and identical events occurring and one rare event matching an existing rare event.

To start, consider randomly tossing a coin twice. The probability of two consecutive heads is 0.25 or

P(heads on the first toss and heads on the second toss) = 0.25.

P(heads on the first toss) × P(heads on the second toss) = 0.5 × 0.5 = 0.25.

The probability a second heads occurs when the first toss is heads is 0.5,

P(heads on the second toss when heads has occurred on the first toss) = 0.5.

In symbols, then $P(heads_2 \mid heads_1) = P(heads_2)$.

The result of the first toss has no influence on the outcome of the second toss because the two tosses are unrelated, referred to as statistically independent outcomes. It is occasionally said "coins have no memory."

These expressions describe statistical independence. That is, the result on the second toss has the same probability regardless of the result on the first toss. If this were not the case, a coin toss would not be used to determine which team kicks off to start a football game.

Lottery results are necessarily independent. That is, random selection of the lottery numbers produces independent results. No predictable patterns occur. The lottery results are then protected from players repeatedly capitalizing on specific patterns to win large amounts of money. If a pattern existed, there would be substantial numbers of repeat winners. Thus, the second six lottery numbers (week two) in the Bulgarian lottery are certainly independent of the previous six numbers selected (week one). In other words, the probability of matching results remains unaffected by previous lottery outcomes. Like tossing a coin, this probability remains the same as the previous week, close to one in 3.4 million.

Remarkably, the following week (week three) an unusual number of players selected the exact same six numbers as the previous two sets of six identical numbers. Perhaps they thought the machine that produced random lottery numbers was broken. It was also incorrectly argued that this was a ridiculous strategy because occurrence of three consecutive sets of six identical lottery numbers is essentially impossible. However, the probability of winning the third lottery is exactly the same as everyone else who bought a lottery ticket regardless of the six numbers chosen.

To repeat, the six lottery numbers selected the following week (week 3) had exactly the same probability as the numbers selected each of the previous two weeks. The occurrence of any six numbers is certainly a rare event. However, the third week numbers were not remarkable in any way and, therefore, unnoticed with the exception of the winner. There are two kinds of rare events: those that are noticed and those that are not. A huge number of unnoticed "rare" events occur every hour of every day. Someone once said, "an extremely rare event would be the occurrence of no rare events."

A headline from a San Francisco newspaper read, "Double Lottery Winner Beats Odds of 1 in 24,000,000,000,000." A Belmont California couple spent 124,000 dollars or 20 dollars a day for 17 years, then hit two jackpots on the same day. The couple won 126,000 dollars from the Fantasy Five lottery and 17 million dollars from the SuperLotto Plus. This is truly a stunning coincidence. However, the fact that they spent 17 years and a considerable amount of money is not relevant to the likelihood of winning twice on a single day. Again, every lottery selection is independent of all previous and future selections. It is worthwhile repeating the words of statistician R. A. Fisher:

"The million, million, million to one chance happens one in a million, million, million times no matter how surprised we may be at the results."

It is notable that the winning couple used a gadget that randomly sorted tiny numbered balls into slots providing their "lucky numbers." Buying multiple tickets for the same lottery or playing two different lotteries does indeed increase chances of winning, but not by much.

When lotteries were first introduced in New York City, a television reporter interviewing an elderly lady asked why she had not yet bought a lottery ticket. Her reply was "the probability of winning is the same whether you buy a ticket or not."

It is traditional to say, "keep trying because sooner or later you will succeed." However, the strategy fails when events are independent because chances of success remain the same regardless of the outcomes of previous independent attempts. However, if a strategy is modified after each failure (creating non-independence), the advise to "keep try-ing" is likely useful.

"Random" is a concept at the heart of statistical methods and analytic results. In addition, random plays important roles in a large variety of important situations. A few examples: encryption, gambling, evolu-tion, and, as noted, lottery outcomes. In encryption, random protects against patterns making deciphering coded material more difficult. In casino gambling, random also guarantees absence of patterns that would be used to increase the probability of winning, something casino owners wish to avoid. In evolution, random distribution of genes in a population is a defense against outside forces capitalizing on patterns that could disrupt successful genetic advantages.

Reference: State Lotteries: How to get in and how to win, by Gail Howard, Ben Buxton (publisher), 1986 (out of print)

22

Fatal coronary disease—risk

Statistics is often about comparisons. It is said when you ask a statistician "How's your wife?", frequently the reply is, "Compared to what?"

A study combining three large quantities of insurance company data ($n = 20{,}995$) reported that among coronary heart disease deaths the % of individuals exposed to at least one or more clinically elevated major risk factors ranged from 87% to 95%. From this study, published in a popular medical journal, the seven authors concluded,

> "antecedent major coronary heart disease (chd) risk factors are common among persons who died from a coronary event, emphasizing the importance of considering all major risk factors in determining risk."

In addition, the authors stated their study suggested an often-made observation that major risk factors are frequently absent in coronary heart disease deaths may be "erroneous."

The study data consisted of only individuals who died from a coronary event. Without comparison to parallel at-risk individuals who did not die, a useful assessment of risk from "exposure to at least one clinically elevated major risk factor" is not possible. For example, among these 20,995 coronary deaths approximately 90% were right-handed and about 10% had a home address ending in the number five. Certainly right-handedness does not increase risk and home address does not decrease risk. More seriously, comparison to the % of individuals exposed to at least one major risk factor among those who did not die is necessary to assess any observed increase in risk. For example, it is possible that individuals who did not die have the similar frequency of exposures observed in the studied group (87% to 95%).

Coronary heart disease data including individuals who did not have a coronary event makes it possible to directly calculate the influence on risk from one or more major coronary risk factors (denoted one^+). That is, these data allow a simple, straightforward, and accurate comparison of risk between those who died and those who did not die. To repeat,

The Joy of Statistics: A Treasury of Elementary Statistical Tools and their Applications. Steve Selvin. © Steve Selvin 2019. Published in 2019 by Oxford University Press. DOI: 10.1093/oso/9780198833444.001.0001

the probability of one or more major risk factors among individuals who died of a coronary event can be a useful number in some contexts but does not directly reflect the influence of risk factors on the likelihood of a coronary event.

Data from an eight-year study consisting of a selected cohort of high risk "middle-class" men aged 40 to 60 years provide an estimate of risk. Unlike the published study based only on *chd*-deaths, individuals with and without *chd*-events are included.

Data: presence/absence of seven risk factors* and occurrence of coronary heart disease events from an eight year study of n = 3153 high risk men:

	chd	no chd	total
one^{+}	246	2373	2619
none	11	523	534
total	257	2896	3153

* = one or more of the major risk factors: height, weight, systolic
blood pressure, diastolic blood pressure, cholesterol level,
smoking exposure, and behavior type—denoted *one^{+}*

Like the study of only *chd*-deaths, the probability of one or more risk factors among individuals with a coronary event is:

$$P(one^{+} \text{ risk factors among men with chd } - events) = 246 / 257 = 0.957.$$

The high occurrence of one or more risk factors is expected because high risk men were deliberately chosen for participation in the study (2619/3153 = 0.831).

The risk of a *chd*-event among men exposed to one or more major risk factors is directly estimated. Thus,

$$P(chd - event \text{ among men with } one^{+} \text{ risk factors}) = 246 / 2619 = 0.094.$$

Conclusion from the published study:

Antecedent major *chd* risk factor exposures were very common among those who developed *chd*, emphasizing the importance of considering all major risk factors in determining risk.

A somewhat relevant comparison:

Occasionally, it is said red cars have the highest risk of being involved in an accident. Again, distinction between two probabilities is important:

$$P(\textit{reported accidents involving red cars}) = P(\text{accident} \mid \text{red})$$

and

$$P(\textit{red cars involved in reported accidents}) = P(\text{red} \mid \text{accident}).$$

The first probability reflects risk. The second probability reflects the frequency of red cars. The proportion of red cars in North America is approximately 10%. The proportion of accidents involving red cars is likely not known.

Perhaps of interest:

Table 22.1 Distribution of car colors in North America

color	percent	color	percent
white	21%	blue	8%
silver	16%	red	10%
black	19%	brown	7%
grey	16%	green	3%

Reference: Personality, type A behavior, and coronary heart disease: The role of emotional expression, Friedman, H. S., and Booth-Kewley, S., *Journal of Personality and Social Psychology*, Vol 53, 1987.

Reference: Greenland P, Knoll MD, Stamler J, et al. Major risk factors as antecedents of fatal and non-fatal coronary. *Journal of the American Medical Association*, 2003; 290(7): 891–7.

23

Pictures

A few unrelated "pictures" illustrate several of the many ways visual representations (graphic plots) can be used to explore statistical issues.

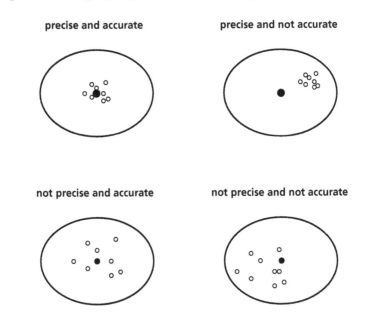

The words accuracy and precision are common in everyday language. For example, "that is an accurate statement, you are precisely right." However, in a statistical context, these two words have separate, unambiguous roles in describing estimates from data.

Accuracy describes the degree an estimated value is correct or "on target." The term bias refers to lack of accuracy. Precision refers to differences among values used to estimate a specific quantity. Low variability yields high precision.

The Joy of Statistics: A Treasury of Elementary Statistical Tools and their Applications. Steve Selvin. © Steve Selvin 2019. Published in 2019 by Oxford University Press. DOI: 10.1093/oso/9780198833444.001.0001

Important distinctions:

When a value is added to or subtracted from sampled observations, it influences accuracy and not precision. When sample size is increased or decreased, it influences precision and not accuracy.

Examples:

The presidential election of 1936 was between candidates Alfred Landon (Republican) and Franklin D. Roosevelt (Democrat). *Literary Digest*, a well-known and popular magazine, traditionally conducted a national survey to predict the outcome of US presidential elections. Previous predictions had a history of being accurate, beginning in 1916.

For the 1936 election, the magazine contacted 10 million potential survey participants using listings from telephone records, magazine subscribers, and rosters from various clubs and organizations. Approximately 2.3 million surveys were answered. From the collected data, the magazine published the prediction that Landon would obtain 57.1% of the popular vote.

If a coin is tossed two million times, the occurrence of heads will be essentially 50% of the tosses. For example, a computer simulated "toss of a fair coin" 2 million times yielded the proportion of heads as 0.50026. The precision of the *Literary Digest* survey, also based on 2 million responses, was thought to be similarly infallible. However, Roosevelt obtained 60.8% of the popular vote and 523 out of 531 possible electoral college votes, one of the largest majorities in history.

What went wrong? The population sampled was unrepresentative of much the US voting public. The year 1936 was the middle of the great depression and many voters could not afford phones or subscriptions to up-scale magazines or club memberships. Thus, precision is always an important goal in statistical estimation but useless without its valuable companion accuracy.

Prior to the 1948 presidential election, the Gallup poll predicted Thomas Dewey would defeat Harry Truman 55% to 44% based on approximately 50,000 survey participants. However, Truman received 50% and Dewey 45% of the popular vote. A famous picture exists of newly elected Truman with a huge smile holding up the morning edition of a Chicago newspaper with headlines "DEWEY DEFEATS TRUMAN."

What went wrong this time? Two mistakes: again a badly chosen strategy for selecting representative survey participants, and ending collection of survey data early in September, a month before the election. The "experts" thought no important further changes in voting trends would occur, an error causing considerable bias. Again, lack of accuracy rendered a likely precise estimate useless.

To predict the 2012 presidential election outcome, the Gallop survey improved but was not yet perfect. A telephone survey of 3117 respondents produced the prediction Mitt Romney would receive 50% and Barack Obama 49% of the popular vote. The election results were essentially the reverse.

Two useful pictures:

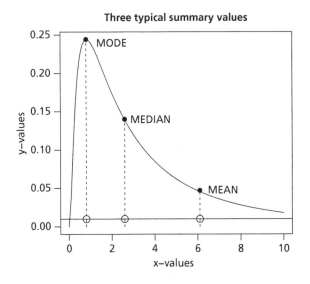

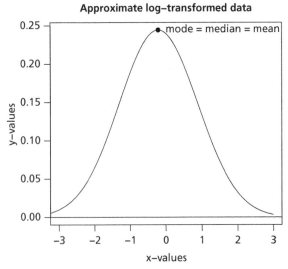

A positive skewed distribution has a left tail displaying large y-values generated by small x-values, called skewed to the right. A negative skewed distribution has a right tail displaying large y-values generated by large x-values, called skewed to the left.

Both plots display the same 20% increase. The first plot is occasionally used because it is compact and esthetically pleasing. However, it is visually misleading.

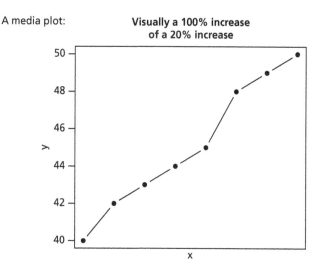

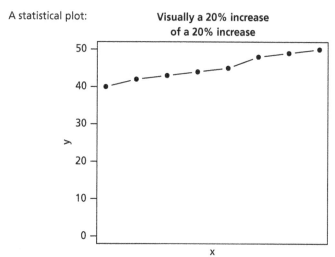

The balance between populations of Canadian lynx and snowshoe hare illustrates a classic and often documented predator/prey relationship.

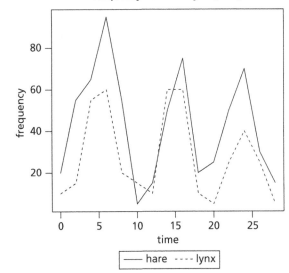

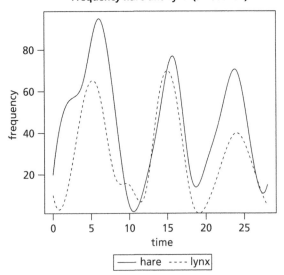

A smoothed plot is typically created by computer graphics or statistical software. The difference between directly plotting data as recorded and a smoothed description is usually a matter of taste. One version is accurate and the other is parsimonious.

Newspaper headlines: "Usain Bolt smashes world record in 100 meter dash." The difference between the new world record time and previous record was 0.11 seconds...smashed???

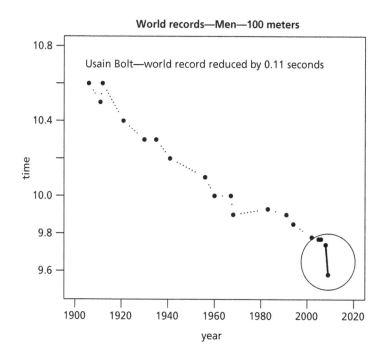

From the picture (internet data), the answer is yes!

Case/control study of childhood cancer:

Table 23.1 Oxford survey of childhood cancer data

x-ray	0	1	2	3	4	5+
cases	7332	287	199	96	59	65
controls	7673	239	154	65	28	29
total	15005	526	353	161	87	94
proportion	0.489	0.546	0.564	0.596	0.678	0.691

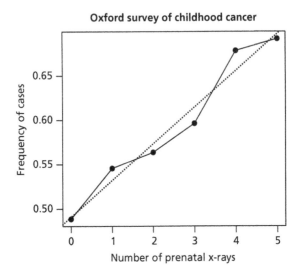

Oxford survey of childhood cancer

In the early 1950s British epidemiologist Alice Stewart and colleagues conducted a case/control study of the relationship between frequency of maternal x-ray exposures during pregnancy and occurrence of childhood cancer. Occasionally, sophisticated and complicated exploration of collected data is not necessary. The Oxford study is definitely such a case. The relationship between x-ray exposure and likelihood of childhood cancer is exceptionally clear. A plot displaying an almost exact linear pattern between number of prenatal x-rays and increase in frequency of childhood cancer unambiguously tells the whole story. Furthermore, a linear increasing relationship is particularly strong evidence of a causal relationship because bias or underlying unmeasured variables or other influences are extremely unlikely to create an almost perfect linear cause/effect pattern by chance. From another perspective, it is especially difficult to envision another explanation likely to produce a clearly linear response. The publication of this case/control study brought a worldwide halt to x-ray as a diagnostic tool during pregnancy. However, it took close to three years to convince the medical community to completely halt x-ray diagnosis of pregnant women.

Statistical model:

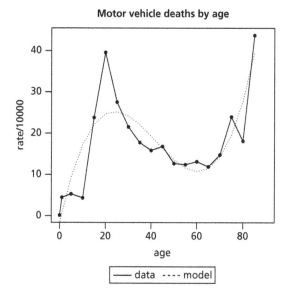

Motor vehicle deaths by age

Table 23.2 Motor vehicle accident deaths

age	0	1–4	5–9	10–14	15–19	20–24	25–29	30–34
at-risk*	1,000,000	987,084	983,740	982,069	982,305	972,751	962,850	953,091
deaths	15	444	524	429	2339	3849	2651	2054

age	35–39	40–44	45–49	50–54	55–59	60–64	65–69	70–74
at-risk*	943,412	932,421	916,982	892,703	856,379	802,800	726,601	621,996
deaths	1673	1476	1541	1133	1064	1055	865	919

age	75–79	80–84	85+
at-risk*	492,026	344,954	199,263
deaths	1180	625	871

* = from relevant life table data

The age/mortality plot describes observed data (solid line) and a mathematical summary (dotted line). Such summary representations of a relationship are typically generated from a statistical model. A pragmatically chosen model describing the pattern of motor vehicle deaths is the cubic equation:

$$estimated\ rate_i = b_0 + b_1 age_i + b_2 age_i^2 + b_3 age_i^3 \quad i = 1, 2, \cdots, 19.$$

Mathematical models often play an important role in a large variety of statistical analyses. They typically provide succinct, compact, and often simple descriptions of typically complicated and almost always unknown relationships. Model summaries usually require complex and extensive calculations. However, modern computer statistical software packages make it possible to easily create and display such model relationships based on observed data, producing sometimes visually simple summary representations of an unknown relationship.

Among statisticians, it is occasionally said, "statistical models are always wrong but some are useful." In other words, statistical models often create useful and clear descriptive summaries but almost never directly identify unknown "true" underlying relationships.

An ugly histogram:

A histogram reveals a distribution of reported cigarettes smoked per day influenced by extreme "digit preference." A sometimes used classification of reported smoking exposure is {none, mild, moderate and extreme}. Clearly these categories poorly reflect the rather chaotic distribution of smoking exposure.

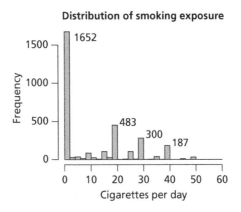

From the example, the category containing 20 cigarettes per day contains at least 483 observations (22% of the data) and the zero category contains 1652 observations (75% of the data). The obvious digit preference produces a distribution where 97% of the data are contained in two of 25 categories. Although no unequivocal strategy exists to create categories for classifying observed values, such as number of cigarettes smoked per day, clearly some care and a bit of finesse is required to create categories that are not misleading.

Birth and death:

Undoubtedly, two important days in a lifetime are day of birth and day of death. Not surprisingly, a number of statistical descriptions and analyses exist of relationships between these two days. For example, an extensive list exists (Internet) of well-known people who died on their birthday. Perhaps the most famous is William Shakespeare.

Documented by several published analyses, a person's day of death and birthday appear slightly associated. From one publication, probability of death on a random day of a year is $1/365 = 0.0027$ and probability of dying on one's birthday was observed to be 0.0031. If day of birth and day of death are unrelated, probability of death on a specific birthday would be the same as the probability of death on a random day. However, a slight positive association appears to exist. In symbols,

$$P(death \,|\, birthday) = 0.0031 \text{ and } P(death) = 0.0027.$$

A sample of $n = 348$ people from a volume of Who's Who in American (1993) allows a graphical exploration of a possible birth/death association. A plot of the relationship between two variables is an effective statistical tool to describe and understand the issues the data were collected to address. For example, a plot can identify meaningful patterns and suggest analytic approaches.

Deaths before and after a birthday:

Table 23.3 Months of deaths classified as before and after a birthday

months	−6	−5	−4	−3	−2	−1	1	2	3	4	5	6	total
deaths	25	23	25	25	27	23	42	40	32	35	30	21	348
expected	29	29	29	29	29	29	29	29	29	29	29	29	348

One of the ten commandments of statistical analysis is "plot your data."

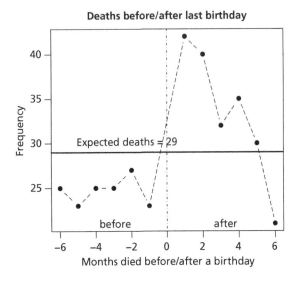

Deaths before/after last birthday

Variation in observed data can be random or systematic. An often useful addition to a plot is a description of the observed values as if their occurrence is exactly random. That is, a complete absence of systematic influences. This artificial conjecture is given the name *null-hypothesis* or sometimes more playfully called the "dull hypothesis." Adding to a plot of these artificial "data" generated as if the observations perfectly conform to the null hypothesis begins to address a key statistical question: Is observed variation likely to be entirely random or does evidence exist of a systematic influence?

When variation in frequency of deaths before/after a birthday is exactly random (null hypothesis), theoretically equal number of deaths would occur each month. That is, if the null hypothesis is exactly true, one twelfth (1/12) of the total deaths would occur each month, ignoring the fact that number of days in each month are not exactly equal. More specifically, from the before/after birthday data, expected number of deaths each month would be 348/12 = 29 (the horizontal line on the plot). A formal comparison of observed data to values based entirely on the conjecture of random variation is often at the center of a statistical analysis. A plot displaying this comparison is a good place to start a more in-depth analysis.

A statistical opposite of random is systematic. However, the difference is not always easy to define or identify. Statistical techniques are specifically created to detect and quantify non-randomness found in collected data. Two simple examples are a graphical introduction.

A circle with radius equal 1.0 and center at $(0, 0)$: random points?

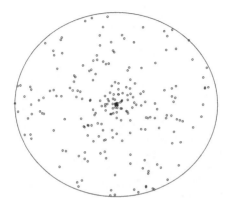

The points ($n = 215$) are clearly not random in a broad sense. They all are contained in a circle. However, the distances measured from the center of the circle to each point are randomly distributed. The 215 distances classified into a table consisting of 15 equal intervals shows close to equal numbers in each interval. These intervals would each contain $215/15 = 15$ distances if the data are exactly "random."

Table 23.4 Random?

	0.2	0.4	0.6	0.8	1.0	1.2	1.4	1.6	1.8	2.0	2.2	2.4	2.6	2.8	3.0
counts	11	13	14	18	15	15	8	11	16	23	20	10	19	15	17
null values	15	15	15	15	15	15	15	15	15	15	15	15	15	15	15

Randomness of the distance of 215 points from the center of the circle is not obvious from the plot. Visually the points appear to cluster around the center. Like many approaches to describing properties of collected data, results depend on the summary statistic chosen. The presence or absence of random values is no exception.

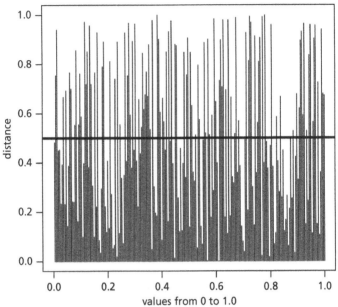

Distance from the circle center for n = 215 points
Mean value = 0.490

One hundred X's and O's:

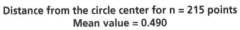

Random Xs and Os????

O X X O O X X X O X O X X X X O O O O X X O X X X

O X O X X X O O X O O O O O X X O O O O O O X O O

O O X O O X X O O O O X O O X O X X X O X X X O X

X O O O X O X X O X X O O O X O O O X X O X X X O

To describe randomness of a distribution of binary values denoted X and O, a definition of the term "run" is necessary. A statistical "run" is defined as the number of identical contiguous elements in a series of binary observations. For example, the sequence {X X O O O} consists of runs of two X-values and three O-values. Again, visual assessment of X/O-randomness is not very satisfactory. The distribution of run lengths of X and O classified into a table frequently creates a useful indication.

Table 23.5 Numbers of each run length

	1	2	3	4	5	6	total
X	9	8	6	1	1	0	25
O	13	4	3	3	1	1	25
total	22	12	9	4	2	1	50

The table hints that the 100 X's and O's may be randomly distributed. The constructed table indicates the occurrences of not very different six run lengths when frequencies of runs of X's are compared to frequencies of runs of O's.

The *Butterfly effect* is the name given to a set of circumstances where small changes in initial conditions amplify into a large chaotic pattern. The theory and applications of this effect are due primarily to physicist Edward Lorenz who described the tiny fluttering of a butterfly's wings amplifying into a hurricane. Although his description is certainly a metaphor, the concept has been applied to improved weather prediction. The plot is a graphical description of a mathematical "Butterfly effect."

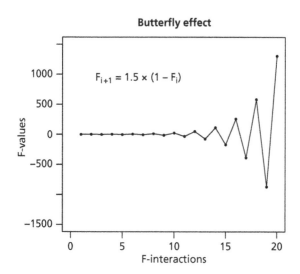

Butterfly effect

$$F_{i+1} = 1.5 \times (1 - F_i)$$

Reference: All data for these plots were directly down-loaded from the Internet with the exception of the last three data sets which were generated by statistical software also found on the Internet.

Reference: Stewart AM, Webb JW, Giles BD, Hewitt D. Preliminary communication: Malignant disease in childhood and diagnostic irradiation In-Utero. *Lancet*, 1956; 2: 447.

Reference: Motor vehicle deaths by age—California state vital records (1998), males. Available at https://www.cdph.ca.gov/Programs/CHSI/Pages/Birth,-Death, -Fetal-Death,-Still-Birth--Marriage-Certificates.aspx (accessed 3 December 2018).

24

The Monty Hall problem

It is Let's Make a Deal—The famous TV show staring Monty Hall (1975).

Monty Hall: One of the three boxes labeled A, B, and C contains the key to the new 1975 Lincoln Continental. The other two boxes are empty. If you choose the box with the key, you win the car.

Contestant: Gasp!

Monty Hall: Select one of the boxes.

Contestant: I'll take box B.

Monty Hall: Now box A and box C are on the table and here is box B (contestant grips box B tightly). It is possible the car key is in that box. I'll give you 100 dollars for the box and the game is over.

Contestant: No, thank you.

Monty Hall: How about 200 dollars.

Audience: No!

Contestant: NO!

Monty Hall: Remember that the probability your box contains the key to the car is 1/3 and the probability it is empty is 2/3. I'll give you 500 dollars.

Audience: No!!

Contestant: No, I think I'll keep this box.

Monty Hall: I'll do you a favor and open one of the remaining boxes on the table (he opens box A). It is empty (Audience: applause). Now either box C on the table or your box B contains the car key. Since two boxes are left, the probability your box contains the key is now 1/2. I'll give you 1000 dollars cash for the box.

Contestant: I'll trade you my box B for your box C on the table.

Monty Hall: That's weird!!!

The Joy of Statistics: A Treasury of Elementary Statistical Tools and their Applications. Steve Selvin. © Steve Selvin 2019. Published in 2019 by Oxford University Press. DOI: 10.1093/oso/9780198833444.001.0001

Question:
Should the contestant trade selected box B for box C on the table????

Two possibilities to consider

After the contestant selects box B, two boxes are left on the table (A and C) and Monty Hall opens box A. Box C remains on the table unopened. There are two possibilities:

1. The probability the key is in the remaining closed box C on the table is the same as the contestant's box B, namely 1/2. There are two boxes and one key. That is, there are two remaining unopened boxes (B and C) and the key equally has to be in one or the other. Knowledge of which of the two original boxes on the table is empty is not relevant since one of the two boxes on the table is necessarily empty.

2. One of the boxes on the table has to be empty and knowing which box changes the probability associated with the remaining closed box. That is, the original probability the key is on the table is 2/3 and when Monty Hall opens the empty box the probability the remaining box contains the key increases from 1/3 to 2/3.

If you chose the first possibility (equal probabilities), you have joined an extremely large number of people who also did not get the correct answer.

Answer 1:

If the contestant switches boxes and if the probability is 1/2 the key is in box C on the table or 1/2 the key is in the contestant's box B, nothing is lost. If the probability is 2/3 the key is in box C on the table, then switching is a good idea. Thus, switching is the best strategy. The question is answered but the problem is not solved.

Answer 2:

Envision 100 boxes on a table, not three. The contestant chooses a single box. The small probability of winning the car is 1/100. When Monty Hall opens boxes on the table the probabilities change. In the most extreme case, Monty opens 98 empty boxes. One closed box remains on the table. This remaining box has a probability of 99/100 = 0.99 of containing the key. Then, switching almost assures a win. When two boxes are on the table the same principle applies. Opening the empty box on the table increases the likelihood that the remaining unopened box on the table contains the key from 1/3 to 2/3.

Answer 3:

List all possible outcomes:

Table 24.1 Enumeration of all possible outcomes

Keys are in box	Contestant chooses box	Monty Hall opens box	Constant switches	Result —
A	A	B or C	A for B or C	loses
A	B	C	B for A	wins
A	C	B	C for A	wins
B	A	C	A for B	wins
B	B	A or C	B for A or C	loses
B	C	A	C for B	wins
C	A	B	A for C	wins
C	B	A	B for C	wins
C	C	A or B	C for A or B	loses

There are nine equally likely possible outcomes. Switching boxes produces six wins and three losses. Therefore, the answer is switch. Listing all possible outcomes is always a good idea when assessing the probability of any event.

Mathematicians sometimes point out probabilities associated with winning the car can not be determined unless it is known whether Monty Hall opens a box on the table at random or he knows which box contains the key and always opens the empty box. Statisticians know that one does not become a multi-millionaire, a famous quiz show host, and inventor of original television game shows for many years without knowing exactly what you are doing.

Of course, the Internet contains details, hundreds of references, considerable debate, and titles of several other books.

```
                          monty      hall
```

May 12, 1975

Mr. Steve Selvin
Asst. Professor of Biostatistics
University of California, Berkeley
Earl Warren Hall
Berkeley, California 94720

Dear Steve:

Thank you for sending me the problem from "The American Statistician."

Although I am not a student of statistics problems, I do know
that these figures can always be used to one's advantage, if I
wished to manipulate same. The big hole in your argument of problems is
that once the first box is seen to be empty, the
contestant cannot exchange his box. So the problems
still remain the same, don't they ... one out of three. Oh, and incidentally,
after one is seen to be empty, his chances are
no longer 50/50 but remain what they were in the first place,
out of three. It just seems to the contestant that one box
having been eliminated, he stands a better chance. Not so. It
was always two to one against him. And if you ever get on my
show, the rules hold fast for you -- no trading boxes after the selection.

Next time let's play on my home grounds. I graduated in
chemistry and zoology. You want to know you chances of
surviving with our polluted air and water?

Sincerely,

Reference: The "whole story" is explored in the book entitled *The Monty Hall Problem* by Jason Rosenhouse, Oxford University Press, 2011.

25

Eye-witness evidence—Collins versus state of California

A woman was robbed in an alley in San Pedro, a city near Los Angeles, California. The police gathered eye-witness evidence at the crime scene. Witnesses reported six identifying characteristics.

Malcolm Collins was found to have all six characteristics and was arrested.

Table 25.1 Probabilities presented at trial

	characteristics	probabilities
1.	Yellow car	1/10
2.	African-American	1/10
3.	Man with a mustache	1/4
4.	A girl accomplice with blond hair	1/3
5.	A girl with a pony-tail	1/10
6.	An inter-racial couple	1/1000

Using these probabilities a local college mathematics professor testified that a random person with the same six characteristics would occur with probability

$$(1/10)(1/10)(1/4)(1/3)(1/10)(1/1000) = 1/12\text{-million}.$$

Based on his calculation, the jury agreed with the prosecution that the probability of a match by chance alone was "beyond a reasonable doubt."

After three years in prison, Malcolm Collins' conviction was overturned. The California supreme court written decision contained two statistically important comments:

The Joy of Statistics: A Treasury of Elementary Statistical Tools and their Applications. Steve Selvin. © Steve Selvin 2019. Published in 2019 by Oxford University Press. DOI: 10.1093/oso/9780198833444.001.0001

1. The probabilities used to calculate the likelihood of matching were of "dubious origin." For example, it is unlikely anyone knows the probability of a girl with a pony tail.
2. Of more importance, the six probabilities are not statistically independent making the resulting summary probability incorrect, considerably incorrect. For example, the occurrence of a blond girl with a pony tail accompanying an African-American man has probability of essentially 1.0 of being an inter-racial couple, not 1/1000.

In general, all persons who stand trial match eye-witness accounts. That is, the probability a defendant matches eye-witness evidence is 1.0 or the trial would not occur. If Malcolm Collins were a white male, he would not have been arrested.

Furthermore, for the sake of an example, suppose 24 million people living in southern California could have committed the crime. This number is unknowable but demonstrates the difficulty of interpreting probabilities derived from eye-witness evidence. If the probability 1/12-million was correct, the prosecuter's probability produces:

$$(24\text{-million people}) \times (1 / 12\text{-million}) = \text{two matching persons}$$

who could have committed the robbery. Thus, only two people exist among 24 million eligible to stand trial. The relevant but unknown probability for the jury to consider is the probability of being guilty among only people who match eye-witness evidence. Everyone else has been eliminated.

From the California supreme court decision:

Mathematics, a veritable sorcerer in our society, while assisting the trier of facts in search of truth, must not cast a spell over him.
                                        Chief Justice Stanley Mosk, California supreme court

Reference: Crim. No. 11176. In Bank. Mar. 11, 1968.
Plaintiff and respondent v. Malcolm Ricardo Collins, defendant and appellant.

26

Probabilities and puzzles

A college basketball game begins with players coming on the court wearing warm-up jackets covering their jersey numbers. One by one they take off their jackets during the pregame warm-up. What is the probability among the first six jersey numbers visible no digit exceeds five?

Answer: probability = 1.0.

Basketball rules require all digits be five or less. This rule allows nonverbal communication of player's numbers among referees using fingers to signal. The left hand indicates the first digit and right hand the second digit.

Most men living in Kansas have more than the mean number of legs—true or false?

Answer: true.

A few men have no legs or one leg making the mean number of legs in Kansas slightly less than two. Therefore, the vast majority of Kansas men are above the mean value.

The other day I was introduced to a man who during our conversation mentioned he is one of two children in his family. I asked if he had a sister. What is the probability his answer was yes?

Answer: probability = 2/3.

There are two possibilities. The two relevant family combinations are female/male or male/male. A female/female pair is not possible. Therefore, the probability of male/male pair is 1/3 and female/male pair is 2/3.

The combination lock on a gate has five buttons numbered 1, 2, 3, 4, and 5. To open the gate three different buttons need to be pressed in a specific order. What is the maximum number of permutations of three buttons that would have to be tried to discover the correct combination?

Answer 1. $5 \times 4 \times 3 = 60$.

Answer 2. Using only buttons with the paint worn off, then $3 \times 2 \times 1 = 6$.

The Joy of Statistics: A Treasury of Elementary Statistical Tools and their Applications. Steve Selvin. © Steve Selvin 2019. Published in 2019 by Oxford University Press. DOI: 10.1093/oso/9780198833444.001.0001

James Fenimore Cooper—*Leatherstocking Tales*

Five volumes written by 19th century author James Fenimore Cooper chronicled the life of a specific character named Hawkeye:

Table 26.1 Five books authored by James Fenimore Cooper

	book title	age*	written**
1.	Deerslayer	21	5 (1841)
2.	Last of the Mohicans	28	2 (1826)
3.	Pathfinder	42	4 (1840)
4.	Pioneer	70	1 (1823)
5.	Prairie	82	3 (1827)

* = age of Hawkeye in each Leatherstocking tale
** = order and year the tales were written

A controversy emerged generated by two literary experts who specialized in the works of author James Fenimore Cooper (Alan Axelrod and Norman Mailer). The issue debated was the order in which the Leather Stocking tales should be read. Mailer recommended the order written and Axelrod defended reading the tales in the order Hawkeye aged. Axelrod suggested Cooper himself was on his side because the corresponding titles are in alphabetical order. Mailer said the ordering was a coincidence.

What is the probability the order was unintentional?

Answer: probability of a unique ordering of five books is 0.0083, that is,

$$\text{ordered by chance} = \frac{1}{5 \times 4 \times 3 \times 2 \times 1} = \frac{1}{120} = 0.0083 \text{ — not likely chance?}$$

More pedestrians using a crosswalk are killed by a car than pedestrians who jaywalk.

Answer: the vast majority of pedestrians use a crosswalk.

Most people injured in an automobile accident are less than a mile from their home.

Same answer: the vast majority of drivers travel at least a mile from their home.

A sometimes suggested strategy for winning a coin toss:

Do not let your opponent make the heads/tails decision because heads is chosen 75% of the time.

Answer: heads occurs with probability 0.5 regardless of opponent's selection!!!

A bedroom drawer contains eight white and eight black socks. Without turning on the bedroom light:

1. What is the minimum number of socks one can pick to guarantee a matching pair?
2. What is the minimum number of socks one can pick to guarantee a matching pair of a selected color?

Answers: 3 and 10 socks.

27

Jokes and quotes

Are statisticians normal?

Three out of every four Americans make up 75% of the population.

Statistics is the art of being certain about being uncertain.

There are three kinds of statisticians: those who can count and those who can not count.

Remember, data is always plural.

A lottery is a tax on people who don't understand statistics.

Yogi Berra: Baseball is 90% mental and half physical.

Sherlock Holmes: Data! Data! he cried impatiently, I cannot make bricks without clay.

Albert Einstein: I am convinced that He (god) does not play dice.

Statistical analysis is not a spectator sport.

Odds are always singular.

A reasonable probability is the only certainty.

Chinese proverb: to be uncertain is uncomfortable, to be certain is ridiculous.

When a statistician talks statistics to a biologist they discuss statistics. When a biologist talks to a statistician they discuss biology but when a statistician talks to a statistician, they talk about women.

The Joy of Statistics: A Treasury of Elementary Statistical Tools and their Applications. Steve Selvin. © Steve Selvin 2019. Published in 2019 by Oxford University Press. DOI: 10.1093/oso/9780198833444.001.0001

It was noted men treat women less severely than women treat men. That is, more than half the women on death row murdered their husbands and a third of the men on death row murdered their wives...[11 versus 1400].

A statistician's wife gave birth to twins. He was delighted. When his minister heard the good news, he said "bring them to church on Sunday to be baptized." "No" replied the new father, "we will baptize one and keep the other as a control."

One day three statisticians were deer hunting in the Maine woods. As they came over a rise, in front of them was a large and magnificent buck. The first statistician raised his gun and fired, missing to the left. Immediately the second statistician fired, missing to the right. The third statistician lowered his gun and exclaimed, "we got him."

One winter night during one of the many German air raids on Moscow during World War II, a distinguished Soviet professor of statistics showed up at the local air-raid shelter. He had never appeared there before. "There are seven million people in Moscow," he used to say. "Why should I expect them to hit me?" His friends were astonished to see him and asked what had changed his mind. "Look," he explained. "there are seven million people in Moscow and one elephant. Last night they got the elephant."

from Against the Gods: The Remarkable Story of Risk
by Peter Bernstein, John Wiley and sons, 1998.

Florence Nightingale: To understand God's thoughts, we must study statistics, for these are a measure of His purpose.

Florence Nightingale (b. 1820) essentially created the profession of nursing. In addition, unlike other Victorians of her time, she accomplished this by collecting large quantities of relevant data. Using tabulated values, creating charts and meaningful summary statistics she described problems and suggested solutions. That is, she was first to use statistical tools to rigorously address medical/health issues producing reliable and convincing evidence, leading to improved cures and treatments. Because she turned numerical data into persuasive evidence, it is reasonable to say she was the first "biostatistician."

William Sealy Grosset (b. 1876) was an employee of Guinness brewery who, as a statistician, used the pseudonym Student because the brewery prohibited any kind of publication by employees to prevent possible disclosure of company secrets. As a "hobby," he developed a statistical assessment of an estimated mean value based on properties of the population sampled. His completely novel statistical strategy opened the door to analysis of small samples of data based on a summary statistic when current analyses typically relied on comparison of values from large samples of sometimes hundreds of observations. His revolutionary approach, today called a t-test, was a major contribution to the beginnings of statistical theory, published in 1906.

There was a young man named Gosset
who one day emerged from a closet.
So around nineteen hundred and three
he invented a distribution called "t."
And, statistics flowed like water from a faucet.

An important table:

	unimportant problems	important problems
accurate data	Guinness records car sales sports/baseball	chemistry biology physics
not accurate data	beauty pageants TV/movie ratings Academy awards	EPIDEMIOLOGY

Monopoly probabilities:

	Property	Probability		Property	Probability
1	Mediterranean	0.0238	21	Kentucky	0.0310
2	Community Chest	0.0211	22	Chance	0.0124
3	Baltic	0.0242	23	Indiana	0.0305
4	Income Tax	0.0260	24	Illinois	0.0355
5	Reading Railroad	0.0332	25	B. and O. Railroad	0.0343
6	Oriental	0.0253	26	Atlantic	0.0301
7	Chance	0.0097	27	Ventnor	0.0299
8	Vermont	0.0257	28	Water Works	0.0315
9	Connecticut	0.0257	29	Marvin Gardens	0.0289
10	Just Visiting	0.0254	30	Go to Jail	—
11	St. Charles Place	0.0303	31	Pacific	0.0299
12	Electric Company	0.0310	32	North Carolina	0.0293
13	States	0.0258	33	Community Chest	0.0264
14	Virginia	0.0288	34	Pennsylvania	0.0279
15	Pennsylvania Railroad	0.0313	35	Short Line Railroad	0.0272
16	St. James Place	0.0318	36	Chance	0.0097
17	Community Chest	0.0272	37	Park Place	0.0244
18	Tennessee	0.0335	38	Luxury Tax	0.0244
19	New York	0.0334	39	Boardwalk	0.0346
20	Free Parking	0.0355	40	Go	0.0346

Jail probabilities: Sent to jail = 0.0444, in jail one turn = 0.0370 and in jail two turns = 0.0308

There was a young man from Cascade
who thought statistics was quite a charade.
He hated the first week of class and more.
He found means, variance and probabilities a bore.
But, hypothesis testing paid for all the effort and fuss he made.

Carl Fredrick Gauss, an outstanding scientist and mathematician of the 18th century is the subject of an often told story. One enterprising individual has found 109 versions of this story. It is not even clear it is true. However, truth is often over rated.

In case you missed it. At the age of seven, as the story goes, Gauss's teacher asked him to calculate the sum of consecutive integers from 1 to 100. He almost immediately replied 5050. His astounded teacher asked how he did it.

In modern terms his method is:

Let S denote the sum,

$$S = 1 + 2 + 3 + 4 + \ldots + 97 + 98 + 99 + 100$$

and equally

$$S = 100 + 99 + 98 + 97 + \ldots + 4 + 3 + 2 + 1$$

therefore

$$S + S = 2S = 101 + 101 + 101 + 101 + \ldots + 101 + 101 + 101 + 101 = 100(101).$$

That is, $2S = 100(101)$. The sum S is then $100(101)/2 = 5050$.
In general, the sum of zero to n consecutive integers is $S = n(n + 1)/2$.

Two branches of mathematics exist:
mathematics that is clear and simple
or
obscure and complicated.

An example of the second kind is:

$$S = 1-1 + 1-1 + 1-1 + 1-1 + 1-1 + 1-1 + 1-1 + 1-1 = 0$$

$$0 + S = 0 + 1-1 + 1-1 + 1-1 + 1-1 + 1-1 + 1-1 + 1-1 + 1 = 0$$

$$0 + 2S = 1 + 0 + 0 + 0 + 0 + 0 + 0 + 0 + 0 + 0 + 0 + 0 + 0 + 0 + 0 + 0 = 1$$

or

$$S = \frac{1}{2} ???$$

News bulletin: meaningless statistics rose 4% this year over last year.

Having given the number of instances respectively in which things are both thus and so, in which they are thus but not so, in which they are so but not thus, and in which they are neither thus nor so, it is required to eliminate the general quantitative relative inherent of the mere thingness of the things and to determine the special quantitative relativity subsisting between the thusness and the soness of the things.

M. H. Doolittle

Exercising the right of occasional suppression and slight modification, it is truly absurd to see how plastic a limited number of observations becomes in the hands of men with preconceived ideas.

Francis Galton (1903)

Sir Francis Galton was born in England 1822 and died in 1911. His career and extensive accomplishments defy description. His numerous major contributions encompassed a wide range: statistics (standard deviation, regression to the mean, bivariate normal distribution, correlation coefficient and more), forensics (fingerprints), eugenics (study of twins, "nature versus nurture"), human genetics (laws of genetics and heredity), religion (the power of prayer), meteorology (weather maps), and numerous other fundamental contributions to science and society. He published a large quantity of research papers and books on these topics and others. Undoubtedly, one of the greatest geniuses of the Victorian era or perhaps any era.

After class a student approached his statistics professor and asked, "Could you answer two questions for me?"
The professor replied, "Yes and what is the second question?"

The following joke has practically nothing to do with statistics. However, an extensive survey determined it is the world's funniest joke (more than 100,000 persons voted).

Sherlock Holmes and Dr. Watson go camping and pitch their tent under the stars. During the night Holmes wakes Watson and says: "Watson look at the stars and tell me what you deduce."

Watson says: "I see millions of stars, and even a few planets like Earth and if there are a few planets like Earth out there, there also might be life."

Holmes replies: "Watson, you idiot. Someone stole our tent."

What do these phrases have in common?
1. Now, Ned I am a maiden won.
2. Nurse, I spy gypsies, run.
3. No lemons, no melon.
4. Madam, I'm Adam.

28

A true life puzzle

A number of years ago the registrar at University of California, Berkeley received a letter critical of university admission of more male than female freshman students. The registrar looked into the situation and found indeed the percentage of males admitted was greater than females. Upon further investigation, however, it was discovered the majority of academic departments admitted more female than male applicants. This apparent reversal appeared not to be possible. As the story goes, the registrar gathered up her admissions data and crossed the street to the statistics department for an explanation.

Table 28.1 Artificial admission data that appear contradictory

departments	males applied	admitted	%	females applied	admitted	%	total
sciences	800	480	60.0	100	70	70.0	900
professional	600	300	50.0	50	30	60.0	650
social sciences	300	120	40.0	600	300	50.0	900
liberal arts	1000	100	10.0	1000	200	20.0	2000
total	2700	1000	37.0	1750	600	34.3	4450

Admissions rates:

$$male\,admission: \quad rate = 1000\,/\,2700 = 0.370$$

$$female\,admission: \quad rate = 600\,/\,1750 = 0.343.$$

However, all four academic departments admitted more women than men (10%). The same admission rates in detail are:

$$male\text{-}rate = \frac{1}{2700}[800(0.6) + 600(0.5) + 300(0.4) + 1000(0.1)] = 0.370$$

$$female\text{-}rate = \frac{1}{1750}[100(0.7) + 50(0.6) + 600(0.5) + 1000(0.2)] = 0.343.$$

The Joy of Statistics: A Treasury of Elementary Statistical Tools and their Applications. Steve Selvin. © Steve Selvin 2019. Published in 2019 by Oxford University Press. DOI: 10.1093/oso/9780198833444.001.0001

These two expressions emphasize the rather different male and female application distributions used to estimate admission rates.

Adjusted admission rates:

$$adjusted\ male-rate = \frac{1}{4450}[900(0.6)+650(0.5)+900(0.4) \\ +2000(0.1)] = 0.320$$

and

$$adjusted\ female-rate = \frac{1}{4450}[900(0.7)+650(0.6) \\ +900(0.5)+2000(0.2)] = 0.420.$$

The male and female adjusted admission rates are calculated using identical application distributions, namely the total applications (last column in the table). Using the same "application distribution" guarantees a comparison between admission rates free from influence of differing male and female application distributions. Equalizing the influence of the admission distributions, the resulting rates, men compared to women, clearly show the expected dominance of admissions of women (32% versus 42%).

This adjustment strategy, often called a *direct adjustment*, is used in a variety of situations. For example, comparing mortality rates between groups usually requires adjustment to accurately compare groups with differing age distributions. Otherwise, like university admissions, an unadjusted comparison of mortality rates produces misleading results.

Actual data from the registrar of the University of California, Berkeley:

Table 28.2 UC admission data

departments	males applied	admitted	%	females applied	admitted	%	total
A	825	512	62.0	108	89	82.0	933
B	520	313	60.0	25	17	68.0	545
C	325	129	40.0	593	202	34.0	918
D	417	138	33.0	375	131	35.0	792
F	191	53	28.0	393	94	24.0	584
G	373	22	6.0	341	24	0.07	714
total	2651	1167	44.0	1835	557	30.4	4486

Admissions:

$$\textit{unadjusted: male admission   rate} = 0.440$$
$$\textit{unadjusted: female admission   rate} = 0.304$$

and

$$\textit{adjusted: male admission   rate} = 0.387$$
$$\textit{adjusted: female admission   rate} = 0.428$$

Again, different distributions of male/female application rates bias comparison between unadjusted admission rates.

Headlines: right-handers outlive lefties by 9-year average

Investigators Halpern and Coren collected 1000 randomly selected death certificates and contacted next of kin to determine left-/right-hand status. They found the average age of death of right-handers was 75 years and left-handers 66 years.

Halpern and Coren: "We were astonished. We had no idea that the difference was going to be this huge."

They attributed the difference to motor vehicle deaths based on the claim that under stress left-handers tend to swerve into oncoming traffic and right-handers away from oncoming traffic.

Artificial data displayed in the following table present another possibility.

Table 28.3 Higher mortality among left-handed individuals???

age	deaths	left	right
25	100	8	92
35	300	21	279
45	600	36	564
55	1000	50	950
65	2000	80	1920
75	3000	90	2910
85	2000	40	1960
95	1000	10	990
sum	10000	335	9665

Mean age at death for right-handed individuals = 71.7 years
Mean age at death for left-handed individuals = 64.5 years
Mean difference: right–left = 7.2 years
Parallel to the university admissions data, the age distributions of right- and left-handed compared individuals are different.
For example,

proportion death age 55 and less — left-handed = 115 / 335 = 0.34

and

proportion death age 55 and less — right-handed = 1885 / 9665 = 0.20.

Like the unadjusted admissions data, it is expected that different distributions produce different mean values.

Reference: Bickel PJ, Hammel EA, O'connell JW. Sex bias in graduate admissions. *Science*, 1975; 187(4175): 398–404

Reference: Coren, S and Halpern DF. Left-handedness: A marker for decreased survival fitness. *Psychological Bulletin*, 1991; 109(1): 90–106.

Reference: Harris, LJ. Do left-handers die sooner than right-handers? Commentary on Coren and Halpern (1991), *Psychological Bulletin*, 114(2): 203–34.

29

Rates—definition and estimation

Many kinds of rates exist. Some examples: interest rates, rates of speed, mortgage rates, and crime rates, to name a few. Two rates often calculated from disease or mortality data are the present topic. An important distinction between these two rates is one reflects likelihood and the other risk. Confusingly, both are usually referred to simply as rates.

A commonly estimated mortality rate is expressed as the number of deaths divided by the number of people who could have died, sometimes called "at-risk" individuals. For example, from California vital records (1998–2007), 231 children died of acute lymphatic leukemia among 3,326,046 who could have died. The rate $231/3,326,046 = 6.9$ deaths per 100,000 at-risk children indicates likelihood of death from leukemia and is essentially an estimated proportion or probability. It is not a measure of risk.

A mortality rate reflecting risk is again based on the number of deaths but accounts for length of time to death, producing a measure of risk. This mortality rate is usually estimated by the number of observed deaths divided by total time lived by the individuals who died.

Artificial examples:

Three deaths in a population of 100 individuals yields a rate of three deaths per 100 at-risk individuals or an estimated probability of death of $3/100 = 0.03$, sometimes said to be 3%. The rate yields a value of 0.03 regardless of whether the deaths occurred in intervals of one year or two years or ten years. The rate remains 0.03.

Again based on 100 individuals, a second rate accounts for the number of deaths and time at risk, such as three deaths in one year, three deaths in two years, and three deaths in ten years. The corresponding rates reflecting risk are: three deaths per 100 person-years (0.03), three deaths per 200 person-years (0.015), and three deaths per 1000 person-years (0.003). Each rate measures risk of death by accounting for total time lived by all individuals who died. Simply, the longer the time lived, the lower the risk.

The Joy of Statistics: A Treasury of Elementary Statistical Tools and their Applications. Steve Selvin. © Steve Selvin 2019. Published in 2019 by Oxford University Press. DOI: 10.1093/oso/9780198833444.001.0001

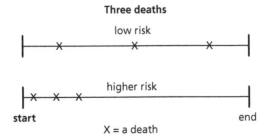

Three deaths

X = a death

Table 29.1 A comparison of two rates

properties		
measures	probability	risk
ranges	$0 \leq q^* \leq 1$	$0 \leq rate \leq$ infinity
units	persons-at-risk	person-years
time alive	not a direct issue	is an issue
life-times	$q^* = 1$	mortality rate $\approx 1/75$

$* = q$ = probability of death

Perhaps the most common rate is speed of a car measured in miles per hour. A property of this rate is the reciprocal value has a useful interpretation. When the child in the back-seat says, "Daddy, are we there yet?" Dad then is concerned with hours per mile. However, if the child asks "Daddy, how fast are we going?," miles per hour is his concern.

Mortality rates have a parallel property. The mean time to death (total person-years alive divided by number of deaths) is the reciprocal of the rate of deaths per person-years (number of deaths divided by total person-years alive).

Artificial examples:

$$mortality\ rate = \frac{10\,deaths}{100\,person - years} = 0.1\,deaths\ per\ person - years.$$

and

$$mean\,time\ alive = \frac{100\,person - years}{10\,deaths} = 10\,person - years\ per\ death.$$

A mortality rate based on person-years is directly calculated when all observed persons have died. To repeat, it is simply the number who died divided by the total person-years lived. For example, when seven individuals each lived {2, 2, 6, 2, 4, 4, 4 years}, the total person-years alive is $2 + 2 + 6 + 2 + 4 + 4 + 4 = 24$ making the mortality rate $= 7$ deaths/24-person-years $= 29.2$ deaths per 100 person-years.

Mortality rates are rarely calculated from data where all at-risk individuals have died because such data rarely exist. Usually, mortality data consist of a mix of individuals, those who have died during a specific time interval yielding "complete" survival times and those who survived beyond the specific time interval under consideration. Individuals who were alive when the observation period ended are said to be *right censored observations*, creating "incomplete" survival times. These individuals are known to be alive at the end of a specific time period but their remaining time alive is not known. The total person-years observed is therefore an underestimate of total person-years lived. That is, a larger value would occur if all at-risk individuals were observed until death. Nevertheless, an approximate estimate of a mortality rate is possible.

To account for "missing" survival times, an assumption, if correct, produces an unbiased estimated mortality rate. The assumption is that all observed individuals have the same constant mean years of remaining lifetime. This assumption is not realistic over a life time or long periods of time. Clearly, older individuals have a shorter mean lifetime than younger individuals. However, for much of human life span, mean survival time is approximately constant, particularly over relatively short time intervals. This property is not only necessary to calculate an unbiased estimated rate, it is implicit when a single rate is estimated from a sample of individuals. A single estimate is only a useful estimate of a specific single and constant rate.

Five examples illustrating constant mean lifetimes. That is, things that do not age or wear out:

1. computer chips (subatomic particles do not deteriorate),
2. airplanes (parts that could fail are regularly replaced),
3. carbon dating of fossils (analysis depends on centuries of constant decay),
4. wine glasses (old wine glasses have the same risk as new glasses of being broken),
5. mortality risk over short periods of time, particularly time periods less than ten years for individuals less than 70 years old.

Perhaps, the mean life-time of wine glasses is two years. However, wine glasses are broken regardless of their age. New glasses will be broken in a few weeks and others will last several years. However, the mean survival time remains two years for all wine glasses regardless of age. In other words, risk of a wine glass being broken does not change with age. A new glass has an expected lifetime of two years and a ten-year-old wine glass has the same two-year expected lifetime. Mortality risk often has approximately the same property, particularly over relatively short periods of time.

How it works:

Notation:

t_i = the observed time lived by the ith individual, censored or complete,

n = number of individuals observed,

d = number of deaths observed,

$n - d$ = number of censored individuals observed, and

μ = constant mean survival time.

The estimated mean survival time lived by all n observed individuals, complete and incomplete, is:

$$\hat{\mu} = \frac{total\ time\ alive}{total\ number\ of\ individuals\ observed} = \frac{\sum t_i + (n-d)\hat{\mu}}{n}$$

making

$$n\hat{\mu} = \sum t_i + n\hat{\mu} - d\hat{\mu}$$

then

$$d\hat{\mu} = \sum t_i \quad \text{and} \quad \hat{\mu} = \frac{\sum t_i}{d} \quad \text{is an estimate of mean survival time.}$$

The value $(n-d)\hat{\mu}$ is an estimate of the unobserved total "missing" time to death lived by $n - d$ individuals with incomplete and constant survival times (μ). Thus,

$$\text{estimated mean survival time} = \hat{\mu} = \frac{\sum t_i}{d} \quad \text{(time/deaths)}$$

and

$$\text{estimated mortality rate} = \frac{d}{\sum t_i} \quad (\text{deaths/time}).$$

Note: when no censored individuals occur, then $\hat{\mu} = \frac{\sum t_i}{n}$. However, when censored individuals are part of the observed data, then $\hat{\mu} = \frac{\sum t_i}{d}$ which compensates for the "missing" incomplete survival times $(d < n)$.

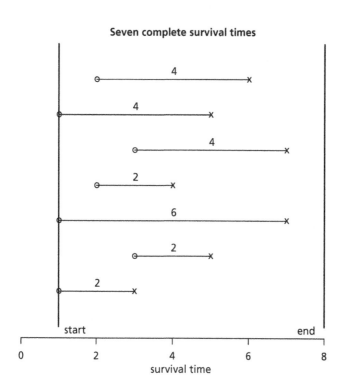

Seven complete survival times

$$\text{estimated mean survival time} = \frac{\sum t_i}{n} = \frac{2+2+6+2+4+4+4}{7} = \frac{24}{7} = 3.43$$

estimated rate $= 1/3.43 = 29.2$ deaths per 100 person-years

Four complete and three incomplete survival times

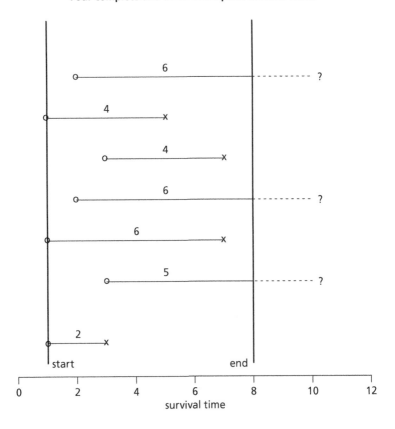

$$\text{estimated mean survival time} = \frac{\sum t_i}{d} = \frac{2+5+6+6+4+4+6}{4} = \frac{33}{4} = 8.25$$

estimated mortality rate $= 1/8.25 = 12.1$ deaths per 100 person-years

30

Geometry of an approximate average rate

Notation:

l_1 = number alive at time t_1

l_2 = number alive at time t_2

$d = l_1 - l_2$ = number died in interval t_1 to t_2

$\delta = t_2 - t_1$ = length of interval t_1 to t_2.

How it works:

$$area = A_1 = \delta l_2 \text{ (rectangle)} \quad \text{and} \quad area = A_2 = \frac{\delta}{2}(l_1 - l_2)\text{ (triangle)}$$

Area = approximate total person-years = rectangle + triangle

$$= A_1 + A_2 = \delta l_2 + \frac{\delta}{2}(l_1 - l_2) = \delta(l_2 + \frac{1}{2}d) = \delta(l_1 - \frac{1}{2}d)$$

where $\delta(l_2 + \frac{1}{2}d)$ or $\delta(l_1 - \frac{1}{2}d)$ estimates total person-years lived between times t_1 to t_2.

Note: $l_1 - l_2 = d$, then $l_1 - \frac{1}{2}d = l_2 + \frac{1}{2}d$.

Geometry of an approximate rate

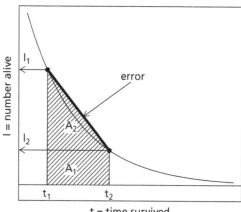

l = number alive

error

A_2

l_1

l_2

A_1

t_1 t_2

t = time survived

The Joy of Statistics: A Treasury of Elementary Statistical Tools and their Applications. Steve Selvin. © Steve Selvin 2019. Published in 2019 by Oxford University Press. DOI: 10.1093/oso/9780198833444.001.0001

Therefore, an approximate average mortality rate is estimated by:

$$estimated\ rate = \hat{R} = \frac{d}{\delta(l_1 - \frac{1}{2}d)}\ person - years.$$

The estimated rate $\hat{R}$ is approximate because a straight line is used to estimate a nonlinear survival curve over a specific time interval. The accuracy of this estimate increases as the length of the interval considered decreases.

In addition, for n persons and d deaths, the approximate average mortality rate is the ratio of two mean values or:

$$estimated\ rate = \hat{R} = \frac{d}{\delta(l_1 - \frac{1}{2}d)} = \frac{d/n}{\delta(l_1 - \frac{1}{2}d)/n}$$

$$= \frac{mean\ number\ of\ deaths}{mean\ time\ alive} = \frac{\overline{d}}{\overline{t}}.$$

One last note:

Incidence rate is the number of new cases recorded over a specific time interval divided by the number of individuals at the start of the time period (a probability).

Prevalence rate is the number of cases of a specific disease at a given time or within a specific interval where the number of these existing cases at a specified time is divided by the total population.

31

Simpson's paradox—two examples and a bit more

When different groups measuring the same relationship are combined they would be expected to yield a single meaningful estimate. Resulting summary values would then be based on increased sample size and a simpler description. That is, combining groups produces fewer and more precise summary values. It is not as simple as it sounds. If groups combined consist of differing relationships, creating a single summary value is rarely useful and can produce just about any kind of meaningless summary value. This phenomenon is frequently referred to as Simpson's paradox. Simpson's paradox is the name given to a situation where groups of data are combined and apparently similar relationships originally observed within each group change dramatically.

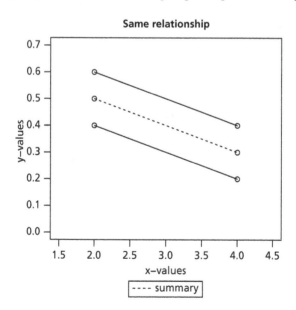

The Joy of Statistics: A Treasury of Elementary Statistical Tools and their Applications. Steve Selvin. © Steve Selvin 2019. Published in 2019 by Oxford University Press. DOI: 10.1093/oso/9780198833444.001.0001

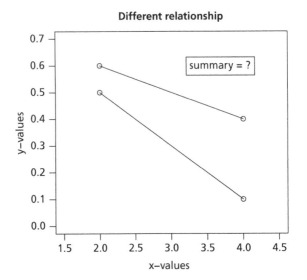

Four plots graphically display underlying issues of a specific instance of Simpson's paradox.

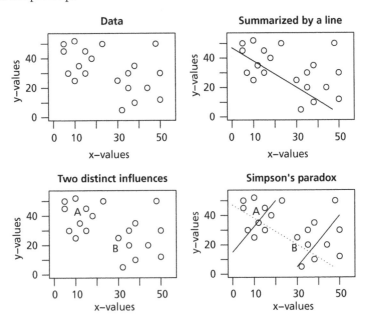

The x/y-data displayed as circles (upper left) are summarized by a straight line (upper right), indicating a negative pattern. However, a third binary variable, labeled A and B, indicates these data consist of two components (lower left), summarized again by a straight lines within each group (lower right). Accounting for the A/B-variable produces an entirely different x/y-relationship. The failure to account for the A/B-variable produces a result often called Simpson's paradox.

Another example—combining 2×2 tables:

Notation:

D = died and $\bar{D}$ = survived

N = new treatment and $\bar{N}$ = usual treatment

Table 31.1 Sample I—$n = 1001$

	N	$\bar{N}$
D	95	800
$\bar{D}$	6	100
	101	900

then, $\widehat{or}_I = 2.0$, $\widehat{or}_{II} = 2.0$ and combined $\widehat{or}_{I+II} = 0.21$.

Table 31.2 Sample II—$n = 1445$

	N	$\bar{N}$
D	400	50
$\bar{D}$	800	195
	1200	245

Table 31.3 Sample I + II—$n = 2446$

	N	$\bar{N}$
D	495	850
$\bar{D}$	806	295
	1301	1145

These artificial data illustrate a pair of 2×2 tables each summarized by an odds ratio. Sample I and Sample II data yield close to the same estimated odds ratio. However, as the example demonstrates, similarity does not guarantee a combined table provides a useful summary value ($\hat{or} = 2$ separate tables versus $\hat{or} = 0.21$ combined table).

A plot of the logarithms of the counts displays the underline complexity of the two odds ratio estimates and illustrates combining data requires care in application and potentially has no useful and sometimes misleading interpretation.

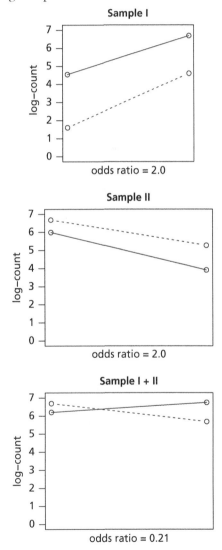

Ecological fallacy

Ecological fallacy is a rather non-descriptive term for the failure of mean or median values or other summary statistics estimated from grouped data to reflect the properties of the observations that make up the groups.

A simple example:

Consider artificial data made up of four groups each containing $n = 5$ observations. Within each group the x/y-relationship is a straight line with negative slope of -10.

Table 31.4 Relationship of individuals within groups

group 1		group 2		group 3		group 4	
x	y	x	y	x	y	x	y
1	50	2	60	3	70	4	80
2	40	3	50	4	60	5	70
3	30	4	40	5	50	6	60
4	20	5	30	6	40	7	50
5	10	6	20	7	30	8	40

The mean values in each group are:

Table 31.5 Relationship among mean values between groups

group	mean x-value	mean y-value
1	$\bar{x}_1 = 3$	$\bar{y}_1 = 30$
2	$\bar{x}_2 = 4$	$\bar{y}_2 = 40$
3	$\bar{x}_3 = 5$	$\bar{y}_3 = 50$
4	$\bar{x}_4 = 6$	$\bar{y}_4 = 60$

The mean values have a linear relationship with a positive slope of $+10$.

Thus, these mean values do not reflect in any way the relationship of the individual observations within the four groups.

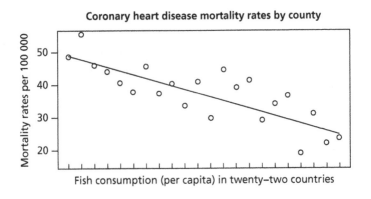

Failure of mean values to reflect within group relationships

The follow questions appeared on the final examination from a graduate course in epidemiology:

True or False: the data indicates an apparent association between increased fish consumption and a lower rate of coronary heat disease mortality.

True or False: the data indicates that increased consumption of fish lowers the risk of coronary heart disease.

It is true that both answers are false.

32

Smoothing—median values

Data values (denoted x) are frequently collected to explore the relationship to another variable (denoted y), motivated by the desire to describe pairwise x/y-relationships. Random variation or other sources of "noise" potentially obscure relationships often revealed by a smoothing strategy. To plot an x/y-relationship, at one extreme is a freehand drawn curve and another extreme is a x/y-description using a postulated mathematical expression. A freehand curve lacks rigor and reproducibility. A mathematically generated curve likely involves unverifiable conjectures. A popular, traditional, and assumption-free compromise is a running median producing an often useful parsimonious description. That is, median values are used to make a series of observations locally more similar to each other creating a smoother, frequently visually simpler, and effective summary description.

One such smoothing technique is achieved by replacing the ith observed value y_i with the median of the three consecutive values $\{y_i-1, y_i, y_i+1\}$ for each observed value x_i ($i = 1, 2, 3, \ldots, n =$ sample size). A "window" of size three is traditional but other choices are not different in principle and might be preferred, depending on circumstances. The choice of an odd number of "window" values guarantees the middle observation is the median when three or more values are ordered from low to high.

An issue immediately arises concerning boundary values of sampled pairs, namely smallest (x_1) and largest (x_n) values. No observations obviously exist below the smallest or above the largest values. A simple solution that allows these first and last observed y-values to be included in a three-value smoothing window is to repeat the values y_1 and y_n, the first and largest values. Three values are then available to select a median value for each of n observations. Other more sophisticated endpoint strategies exist.

Median smoothing—a small example of artificial data:

$$y = \{1, 8, 2, 4, 3\}$$

The Joy of Statistics: A Treasury of Elementary Statistical Tools and their Applications. Steve Selvin. © Steve Selvin 2019. Published in 2019 by Oxford University Press. DOI: 10.1093/oso/9780198833444.001.0001

Data extended by repeating values of first and last end-points:

$$y = \{1,1,8,2,4,3,3\}$$

smoothing:

$$Y_1 = \{\text{median}[\mathbf{1,1,8}],2,4,3,3\} = 1$$
$$Y_2 = \{1,\text{median}[\mathbf{1,8,2}],4,3,3\} = 2$$
$$Y_3 = \{1,1,\text{median}[\mathbf{8,2,4}],3,3\} = 4$$
$$Y_4 = \{1,1,8,\text{median}[\mathbf{2,4,3}],3\} = 3$$
$$Y_5 = \{1,1,8,2,\text{median}[\mathbf{4,3,3}]\} = 3, \text{then,}$$

Smoothed values become: $Y = \{1, 2, 4, 3, 3\}$, graphically:

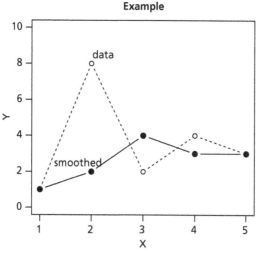

An often useful and sometimes entertaining book entitled *A Handbook of Small Data Sets* edited by D. J. Hand, F. Daly, and several other contributors contains a large number of sometimes strange and unexpected data sets.

Table 32.1 The record of rainfall totals for Sydney
Australia for years 1989 to 2001

year	data	smoothed	residual*
1989	1068	1089	21
1990	909	909	0
1991	841	841	0
1992	475	841	−366
1993	846	475	371
1994	452	846	−394
1995	2830	1397	1433
1996	1397	1397	0
1997	555	988	−433
1998	988	715	273
1999	715	847	−132
2000	847	715	132
2001	509	509	0

* = residual values: difference = smoothed values − data values.

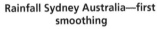

**Rainfall Sydney Australia—first
smoothing**

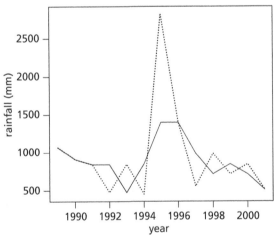

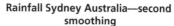

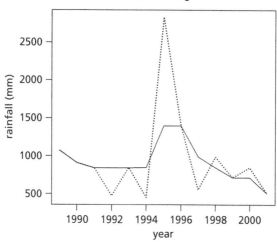

Issues:

Residual values (differences between original data and smoothed values) frequently identify patterns among observed values revealing sometimes unnoticed relationships. The rainfall data is an extreme case, where the year 1995 is likely either an exceptional year or incorrectly recorded. Smoothed values can be further smoothed. Consecutive applications of smoothing applied to already smoothed values can produce stable estimates where further smoothing has no additional effect.

Median smoothing with a "window" of size three is one of a large number of approaches. Some smoothing methods require elaborate computer implementation. Because smoothing simply produces a visual description of a relationship, formally choosing an optimum method is rarely possible. In the case of smoothing, as they say, "What you see, is what you get."

Application:

Several applications illustrate median smoothing applied to Hodgkin's disease mortality data classified by sex and ethnic group.

The obviously large variation in mortality rates by sex and ethnicity is primarily due to the rare occurrence of Hodgkin's disease deaths. That is, small numbers of deaths are highly influenced by incidental variation. As few as two or three additional deaths have a large impact. The age-adjusted US Hodgkin's disease mortality rates per 100,000 persons at risk are: white males = 3.6, white females = 2.6, black males = 3.0, and black females = 1.2.

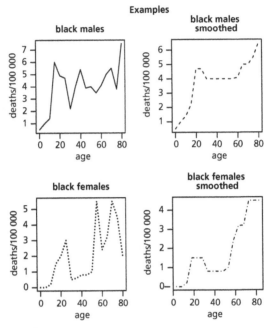

More examples:

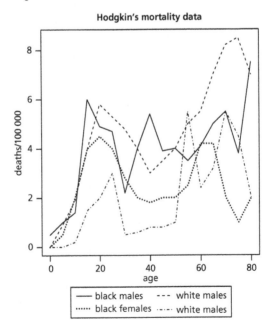

Hodgkin's mortality data—smoothed

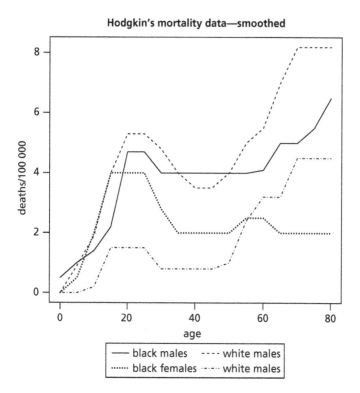

Reference: The Hodgkin's data are an example of the United States government cancer surveillance program.

Note: the Surveillance, Epidemiology, and End Results (SEER) program of the National Cancer Institute provides information and cancer statistics in an effort to reduce cancer risk.

Reference: *A Handbook of Small Data Sets*, Hand, D. J., Daly F., Lunn, A.D., McConway, K.J., and Ostrowski, E., 1994, Chapman and Hall.

33

Two by two table—a missing observation

In an Iowa county, it was important to estimate the number of new-born infants who received a specific vaccine. Two sources of data were available to estimate the total number of vaccinated infants (denoted N). The first source (list 1) was routinely recorded county medical records. The second source (list 2) was a special county-wide survey of newborn infants. When two lists are used to estimate the total number of vaccinated infants, the number of infants not included on either list is not known, technically called a *structural zero*. That is, regardless of the number of families contacted, a count of "missing" infants remains unknown. Nevertheless, the total number of vaccinated infants can be estimated from two lists of data.

Table 33.1 Classification of newborn infants by vaccination status (notation)

		list 2		
		yes	no	total
list 1	yes	a	b	$a + b$
	no	c	$d = ?$	?
	total	$a + c$	?	$N = ?$

To estimate the "missing" count (denoted d) it is necessary that observed infants are located by two independent sources. That is, an infant identified by source 1 with probability denoted p is independently identified by source 2 with probability denoted P. Then, the probability an infant is not included in the collected data is $(1-p) \times (1-P)$. However, the number of "missing" infants $d = N(1-p)(1-P)$ can not be estimated because N is unknown. Nevertheless, a bit of algebra and assumption of

The Joy of Statistics: A Treasury of Elementary Statistical Tools and their Applications. Steve Selvin. © Steve Selvin 2019. Published in 2019 by Oxford University Press. DOI: 10.1093/oso/9780198833444.001.0001

independent sources yield an estimate of the missing count, allowing the 2×2 table to be "completed." That is, total count N can be estimated.

How it works:

In theory, the number of uncounted observations is:

$$d = N(1-p)(1-P) = N\left(\frac{c+d}{N}\right) \times \left(\frac{b+d}{N}\right)$$

where, to repeat, the value d and, therefore, the value N is unknown.

The values of $1-p$ and $1-P$ are:

$$1-p = \frac{c+d}{N} \quad \text{and} \quad 1-P = \frac{b+d}{N}$$

then, in theory, for independent sources,

$$d = N\left[\frac{c+d}{N}\right] \times \left[\frac{b+d}{N}\right] = \frac{1}{N}(bc + cd + bd + d^2) = \frac{1}{N}(bc + d[N-a])$$

making

$$Nd = bc + Nd - ad \quad \text{thus} \quad 0 = bc - ad \quad \text{and} \quad \text{yields} \quad d = \frac{bc}{a}.$$

An estimate of the number of uncounted infants is simply $\hat{d} = \frac{bc}{a}$.

The estimated size of the population sampled becomes $\hat{N} = a+b+c+\hat{d} = a+b+c+\frac{bc}{a}$.

A different version of the identical estimate is $\hat{N} = \frac{(a+c)(a+b)}{a}$.

Table 33.2 An example—Iowa data

		list 2		
		yes	no	total
list 1	yes	$a = 33$	$b = 44$	$a+b = 77$
	no	$c = 14$	$d = ?$	?
	total	$a+c = 47$	?	?

These data yield an estimate $\hat{d} = bc/a = 44(14)/33 = 18.67$ uncounted infants. The county total count is then estimated as $\hat{N} = [33 + 44 + 14] + 18.67 = 109.67$ or 110 vaccinated infants. An estimate from the same data in a different form produces an identical estimated count:

$$\hat{N} = \frac{(a+b)(a+c)}{a} = \frac{(44+33)(33+14)}{33} = \frac{(77)(47)}{33} = 109.67.$$

A short cut:

Independence of two data sources dictates an odds ratio $= or = 1$.

Then, $or = \dfrac{ad}{bc} = 1$ or $ad = bc$ and again $d = \dfrac{bc}{a}$.

The method described to estimate population size from two lists equally applies to estimating size of wild animal populations. A variety of names exist for this estimation strategy. Mark and recapture is perhaps the most popular. Others names are: capture/recapture, mark-release, and sight-resight.

A proportion of a wild animal population is captured, counted, marked, and returned to their environment. A second sample is then collected. Four possibilities are created. Recaptured animals are marked (count $= a$), recaptured animals are unmarked (count $= b$), some marked animals are not recaptured (count $= c$), and finally some animals will not be observed because they were not captured or recaptured—their count (denoted d) is unknown.

Independent counts again provide an estimate of unknown count d making it possible to estimate population size N. As before,

$$\hat{N} = a + b + c + \hat{d} = a + b + c + \frac{bc}{a}.$$

When all animals, captured or not, have the same probability of being captured or recaptured, the observed counts are independent, which is a critical requirement. Some animals are totally unaffected by being caught. For example, Pacific gray whales located at sea off the coast of California and Mexico are tagged with non-injuring markers. The probability of recapture likely remains the same for all gray whales, previously marked or not, creating independence of counts. On the other hand, it was noticed marking and recapturing field mice produced an improbably large estimated population size. Lack of independence was found to be the source of bias. The mice learned non-injuring traps contained food and, once captured and released, these previously captured mice more readily returned to the traps.

Capture/recapture estimation has been applied in an extremely wide variety of situations. A few examples are: to determine the number of sex-workers in Bangkok, Thailand; the number of civilian deaths in war zones; the number of wild dogs in the city of Baltimore; and the number of a rare kind of rabbit that lives in eastern Oregon.

34

Survey data—randomized response

Survey questions occasionally require responses that may not be answered accurately due to reluctance of the person interviewed. A natural reluctance frequently exists to answering personal questions such as: "Do you use recreational drugs?" or "Have you been a victim of sexual abuse?" or "Is your income more than 350,000 dollars per year?" A survey strategy designed to improve cooperation by assuring complete confidentiality is called a *randomization response survey technique.*

This rather clever interview strategy requires a mechanical device-generated response that is a random "yes" or "no" with a known probability (denoted π). When a sensitive yes/no-question is asked, the subject interviewed uses the device to produce a random "yes" or "no" answer and replies to the interviewer only whether the response given by the device is correct or not. Because the interviewer can not see the yes/no-answer given by the device and the subject's response does not reveal the answer to the question, the subject may be less evasive and, perhaps, more truthful. The process guarantees complete confidentiality because the actual answer to the question is not stated or recorded. The collected data are a series of responses indicating only if the subject's answer did or did not agree with the random "answer" generated by the device.

The collected survey data then consist of counts of two outcomes created from four possible responses. The probability the respondent's answer agrees with the random yes/no-answer is $p = P\pi + (1 - P)(1 - \pi)$ or does not agree is $1 - p = P(1 - \pi) + (1 - P)\pi$ where P represents the probability of a true yes-answer and π represents the known probability the device generates a random response "yes." Using predetermined fixed value $\pi = P(random\ "yes\text{-}response")$, the true response P is estimated from n interviews.

The Joy of Statistics: A Treasury of Elementary Statistical Tools and their Applications. Steve Selvin. © Steve Selvin 2019. Published in 2019 by Oxford University Press. DOI: 10.1093/oso/9780198833444.001.0001

Table 34.1 Four probabilities of a randomized response
to an interview question

	subject response		
device	yes	no	total
"yes"	$P\pi$	$(1-P)\pi$	π
"no"	$P(1-\pi)$	$(1-P)(1-\pi)$	$1-\pi$
	P	$1-P$	1.0

An estimate of the proportion of true yes-answers is achieved by equating observed proportion of agreement $\hat{p} = x/n$ to theoretical probability agreement. Symbol x denotes the observed number of n subjects interviewed who stated their answer agreed with the random device-produced yes/no-answer. Specifically, equating data estimate of agreement $\hat{P}$ to corresponding theoretical probability of answer of agreement yields the expression:

$\hat{p} = \dfrac{x}{n} = P\pi + (1-P)(1-\pi)$, and solving for the probability P this expression becomes

$$\hat{P} = \frac{\hat{p} - (1-\pi)}{2\pi - 1}$$

providing an estimate of the frequency of true yes-responses, namely $\hat{P}$.

For example, when $n = 200$ subjects are interviewed and $x = 127$ replied that their answer agreed with the random "answer" given by the device, the observed proportion of agreement is $\hat{p} = 127/200 = 0.635$. Therefore, because π was set at 0.3,

$$\hat{P} = \frac{0.635 - (1-0.3)}{2(0.3)-1} = 0.163$$

estimates the probability P, the true yes-response to the sensitive question.

35

Viral incidence estimation
—a shortcut

It is often important to estimate incidence of a virus. For example, incidence of HIV in a specific group at risk or incidence of West Nile disease virus in a mosquito population or incidence of hepatitis B virus (HBV) in a specific geographic location.

A natural estimate consists of collecting a large number of potentially infected carriers, such as mosquitoes, then testing each individual carrier for presence or absence of virus. This approach is time consuming, can be expensive, and is often subject to error.

An efficient alternative strategy begins with randomly grouping carriers into a series of equal size subsamples. Insects can be directly pooled and biological specimens are equally easy to pool. Each pool is tested and determined to be positive or negative for the presence of virus. A positive pool contains one or more virus-infected carriers and a negative pool contains none.

Notation:

N = number of carriers,
k = number of pools,
n = number of carriers in each pool ($N = nk$),
x = number of pools determined to be negative,
$\hat{Q} = x/k$ is the proportion of negative pools observed among k pools tested,
q = probability of a non-infected carrier, and
$p = 1 - q$ = probability of an infected carrier or incidence.

Instead of N determinations, each of the much smaller number pools, namely k, are determined to be either negative or positive. Incidence is estimated from the observed proportion of negative pools.

The observed proportion of negative pools is $\hat{Q} = x/k$. An expression for the probability of a negative pool is $Q = q^n$. In other words, all n carriers in a single pool are negative. In symbols, this requires:

The Joy of Statistics: A Treasury of Elementary Statistical Tools and their Applications. Steve Selvin. © Steve Selvin 2019. Published in 2019 by Oxford University Press. DOI: 10.1093/oso/9780198833444.001.0001

$$Q = q \times q \times q \times \ldots \times q \times q = q^n.$$

Therefore, an estimated probability of a non-infected carrier $(\hat{q})$ is

$$\hat{q} = \hat{Q}^{1/n} \text{ and } 1 - \hat{q} = \hat{p} = \text{estimated incidence.}$$

For example:

$N = 500, k = 20, n = 25, x = 5$ making $\hat{Q} = x/k = 5/20 = 0.25$, then
$\hat{q} = \hat{Q}^{1/25} = 0.25^{1/25} = 0.946$

and

estimated incidence becomes $\hat{p} = 1 - \hat{q} = 1 - 0.946 = 0.054$

based on $k = 20$ determinations, not $N = 500$.

36

Two-way table—a graphical analysis

A 3 × 4 table of counts of newborn infants with birth defects classified by maternal ethnicity and vitamin use during pregnancy ($n = 411$ mother/infant pairs) illustrates a graphical analysis of a two-way table.

Table 36.1 Data: Vitamin use by ethnicity and estimates

vitamin use	white	African-American	Hispanic	Asian	total
always	50 (55.4)	14 (14.6)	22 (21.0)	39 (34.1)	125
during	84 (72.6)	18 (19.2)	30 (27.5)	32 (44.7)	164
never	48 (54.0)	16 (14.2)	17 (20.5)	41 (33.2)	122
total	182	48	69	112	411

Estimated counts (in parentheses) are calculated as if vitamin use and ethnicity are unrelated (statistically independent). For example, a traditional estimate of a cell count based on the conjecture that ethnicity and vitamin use are unrelated (independent) is:

$$\text{probability} = P(always \text{ and } white) = P(always) \times P(white)$$
$$= \frac{125}{411} \times \frac{182}{411} = 0.135.$$

For these white mothers who always use vitamins (first row and first column), the theoretical count becomes $0.135 \times 411 = 55.4$. The other 11 independent counts are similarly estimated.

Table 36.2 Logarithms of the vitamin/ethnicity data and estimates

	white	African-America	Hispanic	Asian
always	3.91 (4.01)	2.64 (2.68)	3.09 (3.04)	3.66 (3.53)
during	4.43 (4.29)	2.89 (2.95)	3.40 (3.32)	3.47 (3.80)
never	3.87 (3.99)	2.77 (2.66)	2.83 (3.02)	3.71 (3.50)

For example, $\log(50) = 3.91$ and $\log(55.4) = 4.01$.

The Joy of Statistics: A Treasury of Elementary Statistical Tools and their Applications. Steve Selvin. © Steve Selvin 2019. Published in 2019 by Oxford University Press. DOI: 10.1093/oso/9780198833444.001.0001

Table 36.3 Representation of additive data contained in a two-way 3 × 4 table (denoted y_{ij})

	column variable		
row variable	1	2	3
1	$y_{11} = a + r_1 + c_1$	$y_{12} = a + r_1 + c_2$	$y_{13} = a + r_1 + c_3$
2	$y_{21} = a + r_2 + c_1$	$y_{22} = a + r_2 + c_2$	$y_{23} = a + r_2 + c_3$
3	$y_{31} = a + r_3 + c_1$	$y_{32} = a + r_3 + c_2$	$y_{33} = a + r_3 + c_3$
4	$y_{41} = a + r_4 + c_1$	$y_{42} = a + r_4 + c_2$	$y_{43} = a + r_4 + c_3$

* = y_{ij} = logarithm of observed cell count

An additive relationship is key to simple and effective descriptions of data contained in a table. Its defining property is the column variable is unrelated to the row variable. Within all rows, the column values (denoted c) are the same. Within all columns, the row values (denoted r) are the same. Thus, differences between any two columns yield the same value regardless of the row and the differences between any two rows yield the same value regardless of the column. An additive relationship, therefore, dictates exactly no association between row and column variables. To repeat, values c (columns) are the same for all levels of value r (rows) and values r are the same for all levels of value c, when an additive association exists among the logarithms of the table counts.

Two examples:

Difference between row 1 and row 2 is $r_1 - r_2$ for all columns (c has no influence)

and

Difference between column 1 and column 3 is $c_1 - c_3$ for all rows (r has no influence).

Therefore, a plot of an additive relationship of the logarithms of table counts yields parallel lines because row and column variables are unrelated.

Note: $P(AB) = P(A) \times P(B)$, then $log[P(AB)] = log[P(A)] + log[P(B)]$.

Logarithm of independent cell counts (y_{ij}) are similarly additive where

$$log[x_{ij}] = y_{ij} = a + r_i + c_j.$$

A painless graphical analysis of a table

The 3 × 4 table containing counts of newborn infants with a birth defect constructed from three levels of maternal vitamin use (never, during pregnancy, always) and four levels of ethnicity (white, African-American, Hispanic, and Asian) is graphically described by four plots:

1. Logarithms of observed counts (data).
2. Estimated unrelated (additive) row and column log-counts.
3. Both observed and additive log-counts superimposed on the same plot.
4. A specific sensitivity assessment.

Sensitivity assessment is a statistical "experiment" conducted to measure the influence on a specific association by intentionally replacing one or more chosen cell counts to identify the extent to which these values influence the overall row/column relationships. For example, the rather different pattern observed among the 32 Asian women who used vitamins during pregnancy was artificially increased to 42 giving a sense of their influence (row 2 and column 4). Thus, a plot indicates the originally observed difference is primarily due to about 10 Asian women. Sensitivity "experiments" are a simple and general way to identify and explore the degree of influence of a value or values on analytic results. Sensitivity analysis applies not only to tables but is a useful statistical tool in many kinds of analyses.

Log-frequency tables plotted:

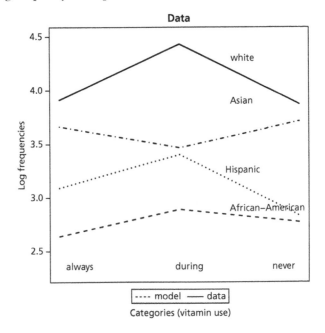

Additive relationship

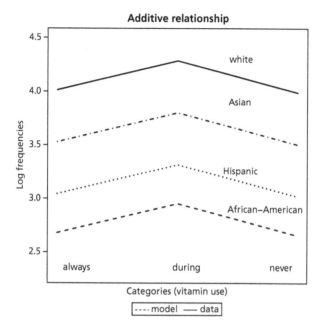

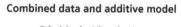

Combined data and additive model

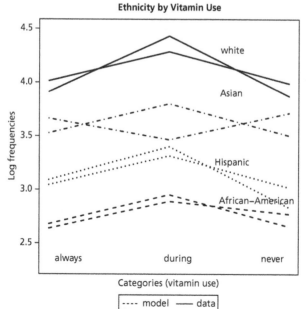

Sensitivity assessment

Ethnicity by Vitamin Use

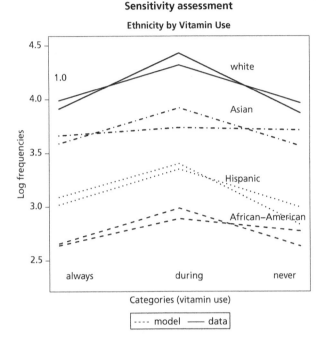

To repeat, to assess sensitivity, the count of Asian mothers who used vitamins during pregnancy was intentionally increased to an artificial 42 from observed value 32 to display the influence of ten specific observations (last plot). Otherwise, ethnicity and vitamin use appear to have separate (independent) influences.

Mean/median polish—an analysis of association within tables

Values distributed into a table, counts or measurements, divide into two meaningful components, named additive and residual values. Or,

observed value = additive value + residual value.

Artificial data illustrate this partition using a process called a *mean polish*.

The first step is to calculate the mean values of each row of the table. For the example table, row mean values are 3.67, 6.00, and 7.67. The three mean values estimate the additive effect in each row. Subtracting these estimates from their respective row values removes their additive influence.

Table 36.4 Example table

	a	b	c
A	1	4	6
B	3	7	8
C	5	9	9

Table 36.5 Example table with row effects removed

	a	b	c
A	−2.66	0.33	2.33
B	−3.00	1.00	2.00
C	−2.66	1.33	1.33

As expected, row values now sum to zero.

The second step is essentially the same as the first but applied to column values of the newly created table. For the example table, column mean values are −2.78, 0.89, and 1.89. Like the row mean values, these column mean values estimate the additive effect of each column variable. Subtracting these estimates from their respective column values removes their additive influence.

Table 36.6 Example table with row and column effects removed

	a	b	c
A	0.11	−0.56	0.44
B	−0.22	0.11	0.12
C	0.11	0.44	−0.56

Like the rows, column values now sum to zero. The remaining values are called *residual values*.

Subtracting residual values from the original data produces a table with exact row/column additive relationships. For example, the observed value in the first row and third column of the example table is six and subtracting the corresponding residual value 0.44 produces the estimated additive value 5.56. That is, the observed count six is partitioned into $6 = 5.56 + 0.44$. Exactly additive estimated values from the example table are:

Table 36.7 Exact additivity

	a	b	c
A	0.89	4.56	5.56
B	3.22	6.89	7.89
C	4.89	8.56	9.56

Thus, *data value = additive value + residual value.*

Residual values directly describe non-additivity between row and column variables in a table. In other words, the location and size of residual values measure association among the variables used to construct the table. For example, if all residual values were zero, row and column variables would be perfectly unrelated (additive). Otherwise, residual values are frequently key to detecting extreme values or identifying informative patterns underlying observed data. In general, residual values are a complete and useful description of extent and pattern of association between row and column variables in a table.

The row variable is parity of newborn infants (birth order) and the column variable is a series of maternal age categories:

Data: table of birth weights (kilograms—New York state birth certificates):

Table 36.8 Maternal age by infant parity

parity	<20	20–25	25–30	30–35	35–40	>40
1	3.28	3.28	3.28	3.22	3.20	3.33
2	3.30	3.35	3.36	3.35	3.33	3.32
3	3.29	3.37	3.39	3.41	3.40	3.40
4	3.26	3.36	3.39	3.42	3.43	3.44
5	3.23	3.32	3.37	3.41	3.44	3.47
6	3.20	3.30	3.34	3.39	3.42	3.48

The row mean values are 3.27, 3.34, 3.38, 3.38, 3.37, and 3.36.

Row additive effects removed:

Table 36.9 Maternal age by infant parity

parity	<20	20–25	25–30	30–35	35–40	>40
1	0.02	0.01	0.01	−0.05	−0.06	0.07
2	−0.03	0.02	0.02	0.01	0.00	−0.02
3	−0.08	0.00	0.01	0.03	0.02	0.02
4	−0.13	−0.03	0.01	0.04	0.05	0.06
5	−0.15	−0.05	0.00	0.04	0.06	0.09
6	−0.15	−0.05	−0.02	0.04	0.06	0.13

The column mean values are −0.09, −0.02, 0.01, 0.02, 0.02, and 0.06.

Column additive effects removed producing 36 residual values:

Table 36.10 Maternal age by infant parity

parity	<20	20–25	25–30	30–35	35–40	>40
1	0.10	0.03	0.01	−0.06	−0.09	0.01
2	0.06	0.04	0.02	−0.01	−0.03	−0.08
3	0.00	0.01	0.01	0.01	0.00	−0.04
4	−0.04	−0.01	0.00	0.02	0.03	0.00
5	−0.06	−0.03	−0.01	0.02	0.04	0.04
6	−0.07	−0.03	−0.03	0.02	0.04	0.07

Note: row and column values add to zero. That is, additive row and column influences are removed leaving 36 residual values.

A graphical display of the pattern of the 36 residual values:

Table 36.11 Maternal age by infant parity

parity	<20	20–25	25–30	30–35	35–40	>40
1	+	+	+	0	0	+
2	+	+	+	0	0	0
3	+	+	+	+	+	0
4	0	0	+	+	+	+
5	0	0	0	+	+	+
6	0	0	0	+	+	+

The pattern of residual values reveals young mothers of high parity infants with reduced birth weights (denoted 0) and similarly older mothers of low parity infants also with reduced birth weights (denoted 0). Thus, the pattern of residual values identifies, in terms of infant birth weight, existence of an optimum parity for each age group.

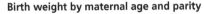

Birth weight by maternal age and parity

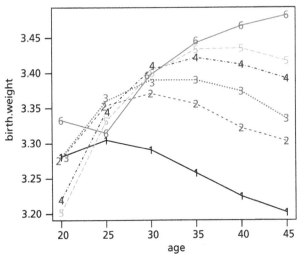

Median polish—another analysis of association in a table

A median polish is not different in principle from the mean polish. The difference is that instead of mean values summarizing row and column additive effects, median values are used. The advantage of using median values is protection against unwanted influences of outlier/extreme values. The disadvantage is estimation of residual values often requires several consecutive iterations to produce stable estimates.

Table 36.12 An example using previous vitamin/ethnicity data

	white	Asian	Hispanic	black
always	50	14	22	39
during	84	18	30	32
never	48	16	17	41

Row median values are 30.5, 31.0, and 29.0.

Table 36.13 Row effects removed

	white	Asian	Hispanic	black
always	19.5	−16.5	−8.5	8.5
during	53.0	−13.0	−1.0	1.0
never	19.0	−13.0	−12.0	12.0

Column median values are 19.5, −13.0, −8.5, and 8.5.

Table 36.14 Column additive effects removed producing residual values

	white	Asian	Hispanic	black
always	0.0	−3.5	0.0	0.0
during	33.5	0.0	7.5	−7.5
never	−0.5	0.0	−3.5	3.5

The table of residual values clearly reveals a single extreme value (infants of white mothers who used vitamins during pregnancy—first column, second row). Otherwise, variables ethnicity and vitamin use appear close to unrelated. That is, the residual values are zero or relatively small. As noted, an additional median polish could be applied to these residual values to "fine tune" the estimated values, yielding residual row and column sums closer to zero (perfect additivity). However, the pattern of residual values is obvious and further analysis would produce essentially the same result.

Table 36.15 Summary relationship—an example table

data values				residual values				additive values		
	a	b	c		a	b	c		a	b	c
A	1	4	6		A	0.11	−0.56	0.44		A	0.89	4.56	5.56
B	3	7	8	−	B	−0.22	0.11	0.12	=	B	3.22	6.89	7.89
C	5	9	9		C	0.11	0.44	−0.56		C	4.89	8.56	9.56

That is, as before, data values − residual values = additive values

Reference: Wasserman CL, Shaw GM, Selvin S, et al. Socioeconomic status, neighborhood conditions, and neural tube defects. *American Journal of Public Health*, 1998; 88(11): 1674–80.

Reference: Selvin S and Garfinkel J. The relationship between parental age and birth order with the percentage of low birth-weight infants. *Human Biology*, 1972; 44(3): 501–9.

37

Data—too good to be true?

From published data:

Table 37.1 Effect of l-carnitine and dl-lipoic acid on the level of GSH, GSSG, GSH/GSSG, and redox index of skeletal muscle of young and aged rats

Particulars	Young control	Young treated	Aged control	Aged treated
GSH	12.68 ± 1.21	12.82 ± 1.26	9.25 ± 0.82	12.53 ± 1.14
GSSG	0.62 ± 0.06	0.505 ± 0.05	1.05 ± 0.09	0.68 ± 0.07
GSH/GSSG	20.68 ± 1.96	23.54 ± 2.35	9.23 ± 1.02	19.54 ± 1.83
Redox index	0.116 ± 0.012	0.128 ± 0.012	0.057 ± 0.005	0.107 ± 0.01

Each value is expressed as a mean ± S.D. for six rats in each group.
Compared with group 1: $P < 0.05$; compared with group 3: $P < 0.05$

Table 37.2 Effect of carnitine and lipoic acid on the activities of GPx, GR, and G6PDH in skeletal muscle of young and aged rats

Particulars	Young control	Young treated	Aged control	Aged treated
GPx	5.34 ± 0.51	5.56 ± 0.56	3.53 ± 0.37	5.16 ± 0.54
GR	0.34 ± 0.03	0.41 ± 0.04	0.23 ± 0.02	0.32 ± 0.03
G6PDH	2.45 ± 0.23	2.58 ± 0.26	1.59 ± 0.17	2.37 ± 0.24

Each value is expressed as a mean ± S.D. for six rats in each group.
Compared with group 1: $P < 0.05$; compared with group 3: $P < 0.05$

Table 37.3 Effect of carnitine and lipoic acid on the level of GSH, GSSG, GSH/GSSG, and redox index in heart of young and aged rats

Particulars	Young control	Young treated	Aged control	Aged treated
GSH	9.42 ± 0.91	9.56 ± 0.94	6.28 ± 0.61	9.02 ± 0.87
GSSG	0.58 ± 0.05	0.51 ± 0.05	0.94 ± 0.08	0.64 ± 0.06
GSH/GSSG	16.31 ± 1.62	18.83 ± 1.84	6.72 ± 0.64	14.12 ± 1.38
Redox index	0.096 ± 0.008	0.107 ± 0.01	0.045 ± 0.004	0.087 ± 0.009

Each value is expressed as a mean ± S.D. for six rats in each group.
Compared with group 1: $P < 0.05$; compared with group 3: $P < 0.05$

The Joy of Statistics: A Treasury of Elementary Statistical Tools and their Applications. Steve Selvin. © Steve Selvin 2019. Published in 2019 by Oxford University Press. DOI: 10.1093/oso/9780198833444.001.0001

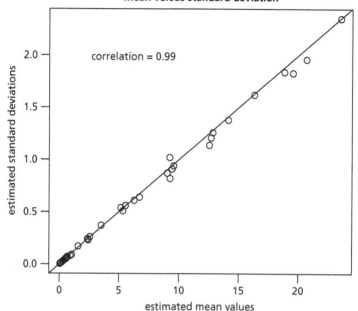

Mean versus standard deviation

correlation = 0.99

(y-axis) estimated standard deviations

(x-axis) estimated mean values

Forty-eight mean values (each of sample size $n = 6$) and their "standard deviations" (denoted S.D.), contained in three tables, have an almost perfect correlation. The correlation coefficient is $r = 0.99$. All 48 mean values divided by 10 produce an estimated "standard deviation" and a statistical test. However, it is not that simple. The mean value divided by ten is not an estimate of the standard deviation and has no meaningful interpretation or use. Peer review of scientific papers is not always a perfect process.

Note: L-carnitine and DL-lipoic acid reverse the age-related deficit in glutathione redox state in skeletal muscle and heart tissues.

Reference: Barbosa Bazotte R and Lopes-Bertolini G. Effects of oral L-carnitine and DL-carnitine supplementation on alloxan-diabetic rats. *Brazilian Archives of Biology and Technology*, 2012.

38

A binary variable—twin pairs

Humans are a diploid organism. Thus, many genetic traits require two alleles (denoted A and a). Genes represented as AA are said to be dominant, Aa said to be heterozygous, and aa said to be recessive. In 1908, G. H. Hardy (a British mathematician) and Wilhelm Weinberg (a German physician) separately discovered alleles are randomly distributed in a population when produced by random mating. Specifically, the dominant AA pair of alleles has frequency p^2 where p represents the frequency of the A-allele. Similarly, the heterozygotic Aa pair of alleles has frequency $2pq$ where $1 - p = q$ represents the frequency of the a-allele and recessive aa pair of alleles has frequency q^2. Because random mating determines many inherited traits not only in humans but also in plants and animals, the Hardy–Weinberg discovery became an early cornerstone in the development of modern genetic theory.

Data—binary allele pairs ($n = 25$), $A = 1$ and $a = 0$ (grouped for convenience of display):

Table 38.1 An artificial example of specific traits resulting from random mating

	AA				Aa												aa								
count	1	2	3	4	5	6	7	8	9	10	11	12	13	14	15	16	17	18	19	20	21	22	23	24	25
$allele_1$	1	1	1	1	0	0	0	0	0	0	1	1	1	1	1	1	0	0	0	0	0	0	0	0	0
$allele_2$	1	1	1	1	1	1	1	1	1	1	0	0	0	0	0	0	0	0	0	0	0	0	0	0	0

Summary:

Table 38.2 Computations

gene type	AA	Aa	aa	sum
binary pair	$\{1, 1\}$	$\{1, 0\}$ or $\{0, 1\}$	$\{0, 0\}$	
frequency	a	b	c	n
probability	p^2	$2pq$	q^2	1.0
count	$a = 4$	$b = 12$	$c = 9$	$n = 25$
probability	0.16	0.48	0.36	1.0

The Joy of Statistics: A Treasury of Elementary Statistical Tools and their Applications. Steve Selvin. © Steve Selvin 2019. Published in 2019 by Oxford University Press. DOI: 10.1093/oso/9780198833444.001.0001

proportion of A-alleles is $p = \dfrac{number\ of\ A - alleles}{total\ number\ of\ alleles} = \dfrac{2a + b}{2n} = 0.4$ and

proportion of a-alleles is $1 - p = q = \dfrac{number\ of\ a - alleles}{total\ number\ of\ alleles} = \dfrac{2c + b}{2n} = 0.6$.

Table 38.3 Hardy–Weinberg equilibrium—random mating
For allele frequency $p = 1 - q$, gene type frequencies are:

random matings	mating frequencies	gene types		
		AA	Aa	aa
$AA \times AA$	p^4	p^4	0	0
$AA \times Aa$	$2p^3q$	p^3q	p^3q	0
$AA \times aa$	p^2q^2	0	p^2q^2	0
$Aa \times AA$	$2p^3q$	p^3q	p^3q	0
$Aa \times Aa$	$4p^2q^2$	p^2q^2	$2p^2q^2$	p^2q^2
$Aa \times aa$	$2pq^3$	0	pq^3	pq^3
$aa \times AA$	p^2q^2	0	p^2q^2	0
$aa \times Aa$	$2pq^3$	0	pq^3	pq^3
$aa \times aa$	q^4	0	0	q^4
generation 2 sum	1.0	p^2	$2pq$	q^2

The A-allele frequency in generation 2 is $p' = p^2 + pq = p$. Thus, random mating produces a population with constant gene frequencies (last row).

The Hardy–Weinberg equilibrium is an elegant and tractable description of the dynamics of a gene pool. Equilibrium results in one generation of random mating from four conditions:

1. all mating is random,
2. a large population,
3. the population is closed—for example, no migration in or out, and
4. mutation or selection or other genetic forces do not change allele frequencies.

When random mating occurs, the frequencies of a large number of genetic relationships directly follow. Some examples are ($p = 0.7$):

Table 38.4 Demonstration: random mating produces a constant A-allele of frequency ($p = 0.7$)

$P(A\text{-allele}) = p = 0.7$

matings	frequencies	AA	Aa	aa
AA × AA	0.24	0.24	0	0
AA × Aa	0.20	0.10	0.10	0
AA × aa	0.05	0	0.05	0
Aa × AA	0.20	0.10	0.10	0
Aa × Aa	0.18	0.05	0.08	0.05
Aa × aa	0.04	0	0.02	0.02
aa × AA	0.05	0	0.05	0
aa × Aa	0.04	0	0.02	0.02
aa × aa	0.01	0	0	0.01
generation 2	1.0	0.49	0.42	0.09

Note: for generation 2, the A-allele frequency

$$p = 0.49 + \frac{1}{2}(0.42) = 0.7 \quad \text{and remains 0.7 for all future}$$

generations of random mating.

Table 38.5 Some probabilities based on random mating

relationships	probability	p = 0.7
aa × aa	q^4	0.01
Aa × Aa	$4p^2q^2$	0.18
$A^+ \times A^+$	$(1 - q^2)^2$	0.83
mother/daughter		
AA and Aa	p^2q	0.15
sib/sib pair		
Aa and Aa	$pq(1 + pq)$	0.25

Note: A^+ denotes gene type of AA or Aa when either type produces the same trait.

Pairs of binary observations are not always randomly distributed. Situations arise where observed pairs consist of a mix of independent and identical binary values. A close look at the statistics of twin pairs in terms of gender illustrates general properties of binary pairs of data.

Twins

Twin births occur in about 1 per 100 live births. Hellin's rule states multiple births occur with probability $(1/100)^{k-1}$ where $k = 2$ for twins, $k = 3$ for triplets and $k = 4$ for quadruplets. Two kinds of twins exist with respect to genetic make-up. Dizygotic twins, sometimes called fraternal twins, whose genetics are the same as sibling pairs (independent inheritance of alleles) and monozygotic twins, sometimes called identical twins, who are genetically identical.

Monozygotic twins in the United States occur at a rate of approximately 3.5 per 1000 live births. This rate is about the same worldwide. Dizygotic twins of young mothers in the United States occur about 6.7 per 1000 live births. This rate approximately doubles for mothers older than 35 years. The frequency of dizygotic twins varies worldwide. Nigeria has the highest rate of dizygotic twins at about 40 per 1000 live births (2010).

Twin pairs can be male/male or male/female or female/female. Using data consisting of twin pairs classified by sex (denoted mm, mf, and ff) produce an estimate of the frequency of monozygotic twins, denoted $\hat{m}$. Male/male twin pairs are both dizygotic (random pairs) and monozygotic twins (identical pairs). The female/female pairs are the same mixture of random and identical twin pairs. The male/female twin pairs are always dizygotic because monozygotic twin pairs are genetically identical and, therefore, are always the same sex. In symbols, where m represents the probability of a monozygotic twin pair and p the probability of a male twin, then $p^2(1 - m)$ is the frequency of dizygotic male/male pairs and pm is the frequency of monozygotic male/male pairs. Thus, frequency of male/male pairs is the mixture $p^2(1 - m) + pm$. The female/female pairs follow the same pattern where $q^2(1 - m) + qm$. However, the male/female pairs, as noted, are always dizygotic and have frequency $2pq(1 - m)$.

Summary:

Binary variables applied to the distribution of twin pairs.

Estimation:

Table 38.6 Expressions for the frequency of twin pairs*

pairs	male/male	male/female	female/female	
frequency	$p^2(1-m) + pm$	$2pq(1-m)$	$q^2(1-m) + qm$	1.0
frequency	$p^2 + pqm$	$2pq - 2pqm$	$q^2 + pqm$	1.0
$m = 0$	p^2	$2pq$	q^2	1.0
$m = 1$	p	0	q	1.0
counts	a	b	c	n

* again m = probability of monozygotic twins.

Table 38.7 An alternative description of twin pair frequencies

	male	female
male	$p^2 + \varepsilon$	$pq - \varepsilon$
female	$pq - \varepsilon$	$q^2 + \varepsilon$

where $\varepsilon = pqm$ represents the deviation from random distribution of twin births. The value ε measures association. That is, perfect independence $\varepsilon = 0$ and perfect association $\varepsilon = pq$.

The total number of males in a data set of twins $(2a + b)$ among n pairs divided by the total number of individuals $(2n)$ again produces the estimated probability of a male twin as:

$$\hat{p} = \frac{2a+b}{2n}.$$

To estimate the probability of monozygotic twins, note the theoretical proportion of male/female pairs is $2pq - 2pqm$. Solving the expression $\frac{b}{n} = 2pq - 2pqm$ (observed frequency of mf-pairs = theoretical frequency of mf-pairs) for the value of m yields estimated monozygotic twin pair frequency $= \hat{m} = 1 - \dfrac{b/n}{2\hat{p}\hat{q}}$

where b/n is the observed frequency of dizygotic twins.

In human populations, the frequency of male and female births is close to $p \approx q \approx 0.5$. Therefore, an alternative estimate of the frequency of monozygotic twin pairs becomes $\hat{m} \approx 1 - 2b/n$, called Weinberg's estimate.

Modern data and more detailed knowledge of the biology of twins shows Weinberg's estimate is a bit biased. However, for an estimate derived at the beginning of the 19th century, it was a brilliant observation.

To illustrate, estimation of the frequency of monozygotic twin pairs is compared between twins with and without a birth defect.

The estimate of the probability of a male twin from the birth defect data is:

$$\hat{p} = \frac{2a+b}{2n} = \frac{2(168)+53}{2(331)} = 0.588$$

Table 38.8 Twins born with and without a birth defect, California (1983–2003)

	mm	mf	ff	total
birth defect	$a = 168$	$b = 53$	$c = 110$	331
no birth defect	$a = 18\ 687$	$b = 16\ 093$	$c = 19\ 188$	53 968

and the estimated probability of a monozygotic twin pair becomes:

$$\hat{m} = 1 - \frac{b}{2n\hat{p}\hat{q}} = 1 - \frac{53}{2(331)(0.588)(0.412)} = 0.670.$$

Weinberg's estimate is $\hat{m} = 1 - \frac{2b}{n} = 1 - \frac{2(53)}{331} = 0.680$. The same estimates for twin pairs without a birth defect are considerably different, where $\hat{p} = 0.495$ and $\hat{m} = 0.405$.

A last note:

Mixtures of random and identical binary pairs occur in a variety of situations and the estimate $\hat{m}$ measures the correlation within these pairs.

An alternative estimate of the correlation coefficient between two binary variables (denoted r) is, for example,

$$r = \frac{4ac - b^2}{(2a+b)(2c+b)} = \hat{m}$$

and from the birth defects twin data:

$$r = \frac{4(168)(110) - 53^2}{(2(168) + 53)(2(110) + 53)} = 0.670 = \hat{m}.$$

The estimate r illustrates one of several versions of a correlation coefficient calculated from two binary variables distributed into pairs. At least five approaches exist to estimate correlation between two binary variables namely: phi, gamma, Pearson, Kendall, and Spearman correlation coefficients. However, all five estimates produce a value identical to the estimate of the frequency of monozygotic twins $\hat{m}$.

Reference: Hardin J, Carmichael SL, Selvin S, et al. Increased prevalence of cardiovascular defects among 53,968 California twin pairs. *American Journal of Medical Genetics*, 2009; 149A(5): 877–86.

39

Mr. Rich and Mr. Poor—a give and take equilibrium

Mr. Rich suggested he would exchange 50% of the money in his wallet for 20% of the money in the wallet of his friend, Mr. Poor. Mr. Poor agreed. Mr. Rich has $100 and Mr. Poor has $10. After the exchange Mr. Rich has $52 and Mr. Poor has $58. Specifically, after the first exchange, Mr. Rich has $100 − $50 + $2 = $52 and Mr. Poor has $10 + $50 − $2 = $58.

Mr. Poor says "let's do that again." After the second exchange, Mr. Rich has $52 − $26 + $11.6 = $37.6 and Mr. Poor has $58 − $11.6 + $26 = $72.4.

Continuing exchanges, Mr. Rich gives less and less money to Mr. Poor while Mr. Poor gives more and more money to Mr. Rich. After ten exchanges, the amount of money Mr. Rich gives Mr. Poor has decreased from $50 to $15.71 and the amount of money Mr. Poor gives Mr. Rich has increased from $2 to $15.71. Additional exchanges of the same amount of money no longer produce change. Technically, it is said that this "give and take" exchange has reached equilibrium.

From a general point of view, for any pair of exchange rates, denoted r_1 for Mr. Rich and r_2 for Mr. Poor, the increase/decrease pattern is described by:

$$\text{Mr. Rich:} \quad R_n = R_{n-1} - [R_{n-1} \times r_1] + [P_{n-1} \times r_2]$$

and

$$\text{Mr. Poor:} \quad P_n = P_{n-1} + [R_{n-1} \times r_1] - [P_{n-1} \times r_2]$$

where R_n represents total amount of Mr. Rich's money and P_n represents the total amount of Mr. Poor's money after the nth exchange.

At equilibrium, amounts exchanged no longer depend on previous exchanges. In symbols, $R_n = R_{n-1} = R$ and $P_n = P_{n-1} = P$. Then, at equilibrium (no change),

$$R = R - [R \times r_1] + [P \times r_2] \quad \text{or} \quad R \times r_1 = P \times r_2,$$

The Joy of Statistics: A Treasury of Elementary Statistical Tools and their Applications. Steve Selvin. © Steve Selvin 2019. Published in 2019 by Oxford University Press. DOI: 10.1093/oso/9780198833444.001.0001

therefore,

$$\text{ratio} = \frac{R}{P} = \frac{r_2}{r_1}$$

regardless of the amount of money at initial exchange.

The basic property of this "give and take" pattern is the ultimate division of money is determined entirely by exchange rates r_1 and r_2. For the example, at equilibrium the initial division of money \$100/\$10 becomes \$31.43/\$78.57 $= 0.4$ determined entirely by the ratio of exchange rates $r_2/r_1 = 0.2/0.5 = 0.4$.

A special case occurs when exchange rates are $r_1 = r$ and $r_2 = 1 - r$. Then

$$R = R - [R \times r] + [P \times (1-r)]$$

then

$$\frac{R}{P} = \frac{1-r}{r}$$

after the first exchange.

A "give and take" pattern is an accurate description of a variety of realistic situations. For example, an important equilibrium occurs when a fatal genetic disease ("take") and a corresponding genetic mutation ("give") establishes an equilibrium created only by selection and mutation rates. Thus, fatal genetic diseases remain at a level entirely determined by rates of selection and mutation.

Table 39.1 Record of exchanges between Mr. R and Mr. P

exchange	exchanges		total after exchange	
	Mr. R to Mr. P	Mr. P to Mr. R	Mr. Rich	Mr. Poor
1.	50.00	2.00	52.00	58.00
2.	26.00	11.60	37.60	72.40
3.	18.80	14.48	33.28	76.72
4.	16.64	15.34	31.98	78.02
5.	15.99	15.60	31.60	78.40
6.	15.80	15.68	31.48	78.52
7.	15.74	15.70	31.44	78.56
8.	15.72	15.71	31.43	78.57
9.	15.72	15.71	31.43	78.57
10.	15.71	15.71	31.43	78.57

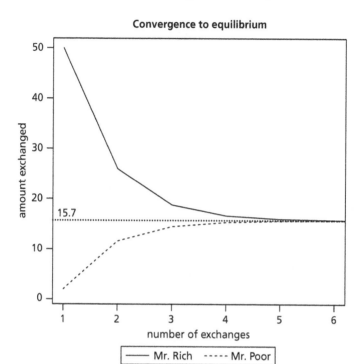

Convergence to equilibrium

40

Log-normal distribution—leukemia and pesticide exposure

A log-normal distribution, sometimes referred to as the "Cinderella" distribution because its older "sister" is the dominant and overbearing famous normal probability distribution*, is an asymmetric distribution. More technically, the logarithms of log-normally distributed values have a normal distribution. Thus, a value denoted Y has a log-normal distribution when a value denoted $X = \log(Y)$ has a normal distribution.

Occasionally small values sampled from an asymmetric log-normal distribution (left tail) are observed but not measured (called *censored values*). These values are often counted but not measured when large values (right tail) are the focus of attention. For example, the description of toxic substance levels.

Estimation of a log-normal distribution and its properties from data containing censored observations is illustrated by four plots. The first plot displays a log-normal distribution with censored values (left tail). The second plot is a logarithmic transformed version of the censored log-normal distribution producing a normal distribution based on the same data with the same censored observations "missing." The third plot displays the corresponding symmetric right tail values from the normal distribution to be used to estimate the corresponding unknown left tail censored values. The fourth plot displays the estimated "complete" normal distribution transformed back (anti-logarithm) to estimate a "complete" log-normal distribution.

---

* The normal distribution is described in every introductory statistics book.

The Joy of Statistics: A Treasury of Elementary Statistical Tools and their Applications. Steve Selvin. © Steve Selvin 2019. Published in 2019 by Oxford University Press. DOI: 10.1093/oso/9780198833444.001.0001

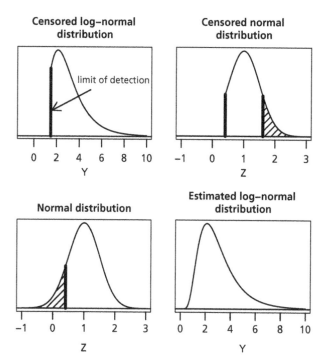

Note: Z = normal distribution and Y = log-normal distribution.

Sophisticated computer-implemented methods exist to estimate a log-normal distribution and its properties from censored data. These methods are optimally precise from a statistical point of view but are not intuitive, require specialized theory for a complete description, and require computer estimation.

How it works:

As suggested, a symmetric distribution right tail can serve as a substitute for a "missing" left tail to produce a "complete distribution." Right tail observations (x_i) have symmetric left tail partners $(\hat{x}_i)$, calculated from the expression $\hat{x}_i = 2m - x_i$. The value m denotes the median value estimated in the usual way.

An illustration:

Example data (x_i): sample size = n = 19, median value m = 10 (in brackets) and censored at $x_7 = 7$ (dots):

$$\text{censored} \qquad \qquad \text{symmetric tail}$$
$$x = \left\{ \ldots .\|7\ 8\ 9\ [\mathbf{10}]\ 11\ 12\ 13\|14\ 15\ 16\ 17\ 18\ 19.\right\}$$

Because $\hat{x} = 2m - x = 20 - x$, then the estimated "missing" values are:

$$\hat{x} = \left\{ 20-19 = 1,\ 20-18 = 2,\ 20-17 = 3,\ 20-16 = 4,\ 20-15 = 5,\ 20-14 = 6 \right\}$$
$$= \left\{ 1, 2, 3, 4, 5, 6 \right\}$$

and the "completed data" become:

$$\text{estimated} \qquad \qquad \text{symmetric tail}$$
$$1\ 2\ 3\ 4\ 5\ 6\|7\ 8\ 9\ [\mathbf{10}]\ 11\ 12\ 13\|14\ 15\ 16\ 17\ 18\ 19$$

The parallel process produces a "complete" normal distribution which can be used to estimate a "complete" log-normal distribution from data with left censored values (small values).

Application:

A sample of $n = 230$ children participated in a case/control study of the herbicide Dacthal as a possible source of risk of childhood leukemia. The collected data contained 76 low Dacthal levels counted but not measured (censored). Using the right tail of the corresponding transformed normal distribution to estimate its left tail values produces a "complete" normal distribution. The estimated mean value $\bar{x}$ and variance S_x^2 from this normal distribution can be used to estimate the corresponding log-normal distribution ($x_i = log[y_i]$). For example, the estimated mean level of Dacthal among cases is $\bar{x} = 0.589$ from the estimated "complete" normal distribution and yields an estimate of the median value of the corresponding log-normal distribution of $e^{0.589} = 1.802$. Other properties of the log-normal distribution are similarly estimated from the mean value and variance of the estimated "complete" normal distribution.

Specifically, the mean value and variance of the normal distribution produce estimates of the summary statistics of the log-normal distribution. For example, from the estimated normal distribution, the estimated mean value $\bar{x} = 0.589$ and variance $S_x^2 = 1.073$ from the leukemia case data yield:

Table 40.1 Relations between the normal and log-normal distributions

	normal	log-normal	estimates*
median	μ_x	e^{α_x}	1.802
mean value	μ_x	$e^{\mu_x + \frac{1}{2}\sigma_x^2}$	3.082
variance	σ_x^2	$\left(e^{2\mu_x + \sigma_x^2}\right)\left(e^{\sigma_x^2} - 1\right)$	18.274
mode	μ_x	$e^{\mu_x - \sigma_x}$	0.640

* = estimates from the case Dacthal data.

Dacthal exposure data estimated distributions

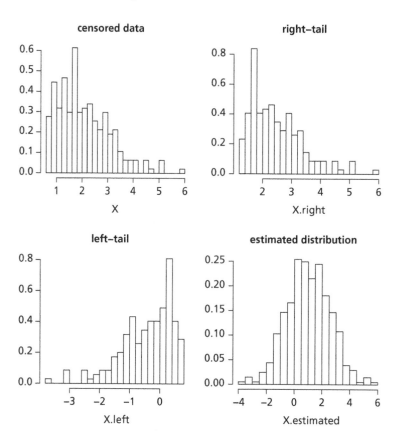

Estimated "complete" log-normal case and control distributions of herbicide Dacthal exposure:

Estimated log–normal distributions

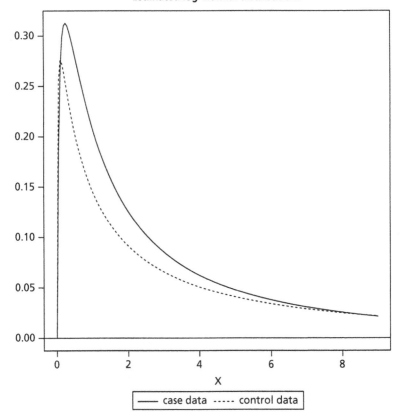

41

A contribution to statistics

by *Wislawa Szymborska (2015)*

Out of a hundred people

those who always know better
—fifty-two,

doubting every step
—nearly all the rest,

glad to lend a hand
if it doesn't take too long
—as high as forty-nine,

always good
because they can't be otherwise
—four, well maybe five,

able to admire without envy
—eighteen,

suffering illusions
induced by fleeting youth
—sixty, give or take a few,

not to be taken lightly
—forty and four,

living in constant fear
of someone or something
—seventy-seven,

capable of happiness
—twenty-something tops,

A Treasury of Elementary Statistical Tools and their Applications. Steve Selvin. © Steve Selvin 2019.
Published in 2019 by Oxford University Press. DOI: 10.1093/oso/9780198833444.001.0001

harmless singly, savage in crowds
—half at least,

cruel
when forced by circumstances
—better not to know
even ball-park figures,

wise after the fact—just a couple more
than wise before it,

taking only things from life
—thirty
(I wish I were wrong),

hunched in pain,
no flashlight in the dark
—eighty-three
sooner or later,

righteous
—thirty-five, which is a lot,

righteous
and understanding
—three,

worthy of compassion
—ninety-nine,

mortal
—a hundred out of a hundred.
Thus far this figure still remains unchanged.

Reference: *Poems New and Old Collected 1957–1997*, by Wislawa Szymborska, translated from Polish by Stanislaw Baranczak and Clare Cavanagh, Hardcourt Inc., 1998

APPENDIX

Golden mean—application of
a quadratic equation

The origin of the "golden mean" occurred around 700 BC. It continues to be of interest from a large variety of perspectives. To list a few:

Architecture, particularly ancient Greek buildings
Painting, particularly Leonardo Da Vinci and Salvador Dali
Music, particularly Debussy and Bartok
Size of everyday credit cards, playing cards, and books
Found in plants, animals, and humans.

Expressed algebraically the golden ratio or golden mean (denoted G) is defined as:

$$\frac{a+b}{a} = \frac{a}{b} = G \quad \text{then,} \quad G = 1.618034\cdots.$$

The geometry of the golden mean

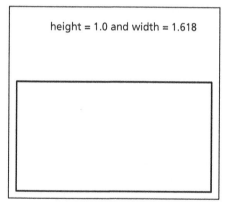

height = 1.0 and width = 1.618

Note: The golden mean rectangle appears in several famous ancient Greek buildings.

A bit of algebra yields the numeric value called the golden mean:

$$\text{definition} = \frac{a+b}{a} = G \quad \text{also,} \quad 1 + \frac{b}{a} = G \quad \text{making} \quad 1 + \frac{1}{G} = G, \quad \text{since } G = \frac{a}{b}.$$

Thus, $G^2 - G - 1 = 0$.

To find the specific value of G:

step 1. $G^2 - G = 1$ adding 1/4 each side of the equation yields

step 2. $G^2 - G + \dfrac{1}{4} = 1 + \dfrac{1}{4}$ or

step 3. $\left(G - \dfrac{1}{2}\right)^2 = \dfrac{5}{4}$ or

step 4. $G - \dfrac{1}{2} = \sqrt{\dfrac{5}{4}}$ and

step 5. $G = \sqrt{\dfrac{5}{4}} + \dfrac{1}{2} = \dfrac{1 + \sqrt{5}}{2} = 1.618034\cdots$.

The same process applied to a general quadratic equation $ax^2 + bx + c$ yields the general solution:

$$x = \frac{-b - \sqrt{b^2 - 4ac}}{2a}.$$

Again, for the golden mean $G^2 - G - 1$, where $a = 1$, $b = -1$ and $c = -1$, then:

$$G = \frac{1 + \sqrt{1 + 4}}{2} = \frac{1 + \sqrt{5}}{2} = 1.618034\cdots$$

Subatomic particles in physics and black holes in astronomy are understood by highly trained physicists and astronomers. For everyone else, it's magic. The phenomenon is understood only by the magicians who know the tricks. The following relationship has such a property.

A well-known infinite sequence of numbers, named after a 12th century Italian scholar, is called a Fibonacci series. This famous sequence is simply generated by adding pairs of consecutive numbers starting at zero. In symbols, $f[k + 2] = f[k] + f[k + 1]$ where $k = 0, 1, 2, 3, \cdots$. Specifically, the Fibonacci sequence is:

$$\left\{\, 0, 1, 1, 2, 3, 5, 8, 13, 21, [\mathbf{34, 55, 89}\,]144, 233, 377, \cdots, \right\}$$

For example, when $k = 10$, then $f[12] = f[10] + f[11] = 34 + 55 = 89$ (brackets). The Fibonacci series, like the golden mean, has been a topic of interest for centuries.

A remarkable relationship exists between the Fibonacci series and the golden mean. The ratio of consecutive Fibonacci values (*ratio* $= f[k]/f[k - 1]$) converges to the golden mean as the number of values in the sequence increase. A few ratios illustrate this magic.

A few converging values from a Fibonacci series

k	3	6	10	15	30
ratio	1/1	5/3	34/21	377/233	514229/317811
ratio	1.000000	1.666667	1.619048	1.618025	1.618034

The chord theorem

The chord theorem states, again "magically," that distance a multiplied by distance d always equals the distance b multiplied by distance c regardless where the chords intersect.

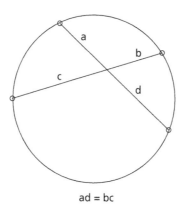

ad = bc

Two intersecting chords present a graphical representation of an odds ratio $= ad/bc = 1$ or independent values in a 2×2 table, but it is in fact not much use. The logic of this relationship arises not from the properties of a circle or chords of a circle but from the geometry of congruent triangles.

Pythagorean theorem—a proof

Relevant geometry: two squares

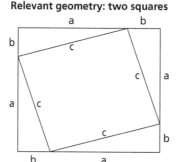

Ninth graders know the Pythagorean theorem. However, few adults can explain why indeed the square of the hypotenuse of a right-angled triangle is equal to the sum of the squares of the two remaining sides.

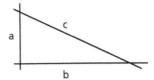

That is, the length of side a squared + the length of side b squared equals the length of side c squared (hypotenuse). In symbols,

$$a^2 + b^2 = c^2.$$

From the geometry:

The area of four triangles plus the area of the square enclosed by the larger square equals the total area of the large square $(a + b) \times (a + b)$ or

$$(a+b)^2 = area\,of\,the\,four\,triangles\left(area = 4 \times \frac{1}{2}ab\right)$$
$$+ area\,of\,the\,enclosed\,square\,(area = c^2).$$

Thus,

$$(a+b)^2 = a^2 + 2ab + b^2 = 4 \times area\,of\,triangles + area\,of\,enclosed\,square =$$
$$4 \times \frac{1}{2}ab + c^2 = 2ab + c^2$$

then

$$(a+b)^2 = a^2 + 2ab + b^2 = 2ab + c^2, \quad yielding\ a^2 + b^2 = c^2$$

and, as they say, QED.

Note: *Quod Erat Demonstrandum* is Latin for "that which was to be demonstrated."

Pi—a famous number

The never ending value π (a circle's circumference divided by its diameter) plays only a small role in statistics but certainly ranks high in the hall of fame of numbers.

A value of π is

$$\pi = 3.14159265358979323846426433\cdots$$

The numeric value of π was first calculated by Archimedes in 212 BC. It is not clear how it was calculated. Ever since, the calculation of π has been one of the most enduring endeavors. Century after century, mathematicians produced more and more accurate values (more decimal places). In 1844, Zacharias Dase produced a value with 200 decimal places in less than two months. This pattern continued until recently when the modern computer spoiled the fun.

The sum of a large number of infinite series produce the value π. Two simple examples among many are:

$$\pi = 4 \times \left[1 + \frac{1}{3} + \frac{1}{5} + \frac{1}{7} + \frac{1}{9} + \frac{1}{11}\cdots \right]$$

and

$$\pi = 2 \times \left[\frac{2 \times 2 \times 4 \times 4 \times 6 \times 6 \times 8 \times 8 \cdots}{1 \times 1 \times 3 \times 3 \times 5 \times 5 \times 7 \times 7 \times 9 \times 9 \cdots} \right].$$

These series converge beyond a few decimal places extremely slowly. The two examples, consisting of 10,000 terms in the series, produce values accurate to only the fourth decimal place. Specifically, the values are 3.14151 and 3.14149 compared to 3.14159....

Another endeavor applied to the value π is the creation of approximate expressions. A few simple examples:

$$\frac{22}{7} = 3.143,$$

$$\sqrt{2} + \sqrt{3} = 3.146,$$

$$\frac{9}{5} + \sqrt{\frac{9}{5}} = 3.1416,$$

and

$$\frac{355}{113} = 3.14159.$$

Spectacularly,

$$\frac{log(640320^3 + 744)}{\sqrt{163}} = 3.14159\cdots \text{ and is exact to 30 decimal places.}$$

Sum of an infinite series

Early Greek mathematicians deduced the sum of an infinite series of numbers could not be exactly determined. This conjecture sounds reasonable since regardless the numbers included in the sum, there remains an infinity of numbers not included. Centuries later, for the infinite sum denoted S $(0 < p < 1)$:

$$S = 1 + p + p^2 + p^3 + p^4 + p^5 + p^6 + \cdots$$

making

$$pS = p + p^2 + p^3 + p^4 + p^5 + p^6 + \cdots$$

then,

$$S - pS = 1 \quad \text{or} \quad S(1-p) = 1 \quad \text{and} \quad S = \frac{p}{1-p}.$$

Therefore, for example, $p = 1/2$, then

$$S = (1/2) + (1/2)^2 + (1/2)^3 + (1/2)^4 + \cdots$$
$$S = 1/2 + 1/4 + 1/8 + 1/16 + 1/32 + \cdots$$

and

$$S = \frac{1/2}{1 - 1/2} = 1.0.$$

Subject Index

Tables and figures are indicated by an italic *t* and *f* following the page number.

Printed and bound by CPI Group (UK) Ltd, Croydon, CR0 4YY